#4

PLATINUM BY CARTIER
TRIUMPHS OF THE JEWELERS' ART

PLATINUM BY CARTIER

TRIUMPHS OF THE JEWELERS'ART

By Franco Cologni and Eric Nussbaum

Harry N. Abrams, Inc., Publishers

with the collaboration of Ezio Sinigaglia

Editorial and artistic director: Ettore Mocchetti
Coordination: Publiprom, Grazia Valtorta
Design: Walter Del Frate, Daniele Cipolat

For Harry N. Abrams, Inc.:
Editor: Ruth A. Peltason
Translator: Lory Frankel

Acknowledgments to:
Michel Aliaga, Patrizia Baini, Georges de Bartha, David Bennett, Philippe Bessis,
Karine Breuninger, Teresa Buxton, Jeremy Coombes, David Cullen,
François Curiel, Arlette Dahan, Simon de Pury, Ralph Destino, Dominique Fléchon,
Charles-Henri Garnier, Lionel Giraud, Anne Holbach, Betty Jais, Micheline Kanouï,
Albert Kaufmann, Guy Leymarie, Carol Logan, Evelyne Lepeu, Pascale Milhaud,
Corentin Quideau, Pierre Rainero, Leslie Roskind, Véronique Sacuto, Bonnie Selfe,
Dominique Sensarric, Christiane Turlure, Christine Urfer, Eric Valdieu,
Wilma Viganò Pandiani, Peter von Zahn, Dieter Zimmermann

Orginally published in France under the title *Cartier: Joaillier du Platine*

Library of Congress Cataloging-in-Publication Data
Cologni, Franco.
    Platinum by Cartier / Franco Cologni, Eric Nussbaum.
        p. cm.
    Includes bibliographical references and index.
    ISBN 0–8109–3738–7 (cloth)
    1. Cartier (Firm) 2. Platinum jewelry–France–History–20th
century. I. Nussbaum, Eric.   II. Title.
    NK7398.C37C66   1996
    739.27′092′2–dc20   95–41868

Typeset by Graphic Service, Milano
Printed by Grafiche Milani, Segrate-Milano
Bound by Piolini e C., Milano

CONTENTS

After exciting the interest of scientists, platinum won over jewelers. Cartier, who wove it in very fine threads to enhance the brilliance of diamonds, contributed to its legitimacy as an irreplaceable precious metal. As the light of the Belle Époque died, it set off sparks that would become legendary.

The Fathers of Platinum page 19

The style characterized by the curved line evolved into new geometric forms that influenced all the decorative arts of the period. Cartier, which had forecast Art Deco in its work, endowed it with a sense of balance and an unequaled originality. In the domain of jewelry, platinum catalyzed this creative evolution, always with the aim of conferring ever more splendor and beauty on precious gems.

The Sources of Platinum page 31

While the investigation of abstract forms led to the sublime perfection of white Art Deco, color and life exploded in the tutti frutti style and in the return to animal themes. The platinum of Cartier's jewels united East and West and color and form in a synthesis of pure art and lasting beauty.

The Properties of Platinum page 43

Among the cultural disarray that followed the war, Cartier rediscovered his most original approach in a fresh and brilliant naturalism. Platinum, partly replaced by gold in the jewels of fantasy, appeared as the king of the night.

The Uses of Platinum page 55

With the success of Les Must, Cartier was renewed. As the 1990s began, platinum came to the fore not only in the area of fine jewelry but also in its new lines of New Jewelry and, especially, of watches.

Preface

Since the publication of Hans Nadelhoffer's book more than ten years ago, a veritable "literature" on Cartier has evolved, detailing the company's legend, history, and creations. In this specialized library, each book holds a particular place and answers a particular need. While Cartier the Legend *by Gilberte Gautier focuses on the great figures of the family who began the company, three other books each cover a specific area of the company's production: jewels (Cartier: Jewelers Extraordinary, by Nadelhoffer), watches* Le Temps de Cartier, *by Jader Barracca, Giampiero Negretti, and Franco Nencini, and accessories and objects of daily use (Made by Cartier, by Franco Cologni and Ettore Mocchetti).*

This book, Platinum by Cartier, *offers a new, original approach that differentiates it from the others. All areas of the company's production and styles it has created are found here. The book's major characters are the Cartier style and platinum, which naturally find their highest expression in fine jewelry. However, many examples of watches, accessories, and objects of all kinds are also represented. One could say that Cartier, having realized with the three preceding monographs the "historic catalogue" of its production, now begins a deeper probing, moving its focus from forms to materials and, at the same time, from results to projects, from effects to causes, from realizations to their underlying culture and art.*

Cartier unquestionably played a crucial and decisive role in the fortunes of platinum in jewelry, which justifies the linking of the jeweler and the material in the title of this work and in the course of the history it narrates. However, as the reader is doubtless aware, neither Cartier nor platinum is the principal actor. More than the jeweler or the precious metal that has so greatly contributed to the spread of its image and its legend, the spotlight falls on the culture of our century, which runs a convoluted course from conformity to rebellion, from old to new, from rationality to irrationality, from hope to tragedy. At an extraordinarily opportune moment, platinum entered the history of jewelry at the beginning of the twentieth century, and Cartier, in the same period, took the road of success and entered into legend. This coincidence of timing necessitates transposing the motif of this volume into an unusual and astonishingly fertile key—that of exploring the connections between these events and the contemporary world and its prodigious creativity, complex and contradictory.

Telling the story of a century in the "art of platinum" presupposes, as experience has shown, the same kind of interdisciplinary investigations as those required by a century of poetry or painting. With its remarkable resistance and ductility, platinum threads politics and modes of living, the decorative arts and fashion, literature and music, the plastic arts and architecture, paying homage not so much to Cartier as to the noble minor art of working with metal and precious gems and to the master artisans who are the interpreters of these materials. Perhaps the most original aspect of this book resides in its desire to elevate the artistic crafts to the level of a true art.

In line with this project, both modest and ambitious, this book has an unusual structure, composed of two parts, text and images, symmetrical but completely separated, in order to preserve their respective identities. Ideally, this will promote equally the readability of the text, setting off its literary and conceptual qualities, and the visual presentation of the images and the pleasure that they can provide: the history of Cartier and of platinum, not in words but in colors and shapes.

Franco Cologni and Eric Nussbaum

THE LEGEND OF PLATINUM

An Egyptian casket dating from 720 B.C. and decorated with hieroglyphics in platinum.

*T*oday, the word platinum *evokes the image of rare, exquisite, and precious objects. Yet its etymology betrays a paradoxical ambiguity. It was the Spanish conquistadors who gave the name to a shiny white metal found in abundance in the gold-bearing sand of areas in New Granada, the Spanish colony in South America corresponding to today's Colombia. Looking for gold, the conquistadors considered this unknown metal an intrusive element, an annoying nuisance. It more or less resembled silver, although it was much heavier, harder, and stronger, resistant even to fire. However, in their understandable ignorance, the Spanish thought silver an infinitely more precious metal. They therefore christened this "lesser silver"* platina, *diminutive of* plata, *the Spanish word for silver. This diminutive, given in contempt rather than affection, has remained the designation of the most precious metal on the planet.*

Platinum was thus a late discovery, its history not truly beginning until the middle of the eighteenth century. Well before then, though, it left isolated, and mysterious, traces. The oldest pieces have come down from the Egyptians, a people who exemplified mystery. It has been shown that some jewels that belonged to the Pharaohs of the Eighteenth Dynasty (1551–1306 B.C.) were made of an alloy of gold and platinum. It is highly doubtful that the artisans who fashioned them were aware of the specific nature of the gold they used. More likely, the gold taken from the Nubian mines worked by the Pharaohs at the time had variable quantities of platinum naturally mixed in, usually a small amount but sometimes, randomly, in significant proportions.

A better known discovery was a casket, also Egyptian, that belonged to the great priestess Shepenupet I, daughter of the king of Thebes, that dates back to the eighth century B.C. This casket is embellished on one side by hieroglyphics in gold and on another by hieroglyphics in silver. It appears, however, that one of these is actually made of platinum, or, rather, platinum alloyed with other metals of the same group. In any case, there is again no way to determine if the platinum was consciously selected by the artisan.

Several small objects discovered at the turn of the century in Ecuador, in the area around Esmeraldas, paint a different picture. These finger rings, lip or nose rings, charms, and a miniature ingot, dating from between the first and fourth centuries A.D., were made of nearly pure platinum (more than 80 percent). They at-

test to a way of working platinum that is unique in the history of ancient civilizations and that was not acquired by Europe until the second half of the eighteenth century. Unlike the Egyptians, the Incas were familiar with platinum, had learned to mine it and work it, and considered it a precious metal, in contrast to their Spanish conquerors. Why such an awareness of platinum should be limited to an extremely narrow region in Ecuador and then shrouded in obscurity for at least thirteen centuries remains a mystery.

Mystery and myth long surrounded platinum even after its discovery in the modern era. For at least two centuries, from the sixteenth to the eighteenth, not a single specimen of the new metal reached Europe, although rumors traveled regarding its exceptional qualities. The scientists of the time were most fascinated by its extremely high melting point, which made it almost impossible to melt.

Meanwhile, the Spanish, in their obsessive quest for gold, continued to come upon this undesirable "little silver." Far from rejoicing, they complained bitterly about it, going so far as to abandon mines that proved excessively "infested" with this cursed element. The platinum could not be removed from the gold with a few blows of the pick or by hammering the mineral on a steel anvil. Abundant, heavy, and mixed with gold in its natural state, probably in the form of dust, platinum was used by unscrupulous merchants to adulterate gold. In response, the Spanish government banned its use and confiscated it where it was mined, scattering it immediately in rivers or streams. For this reason, the first samples of platinum reached Europe secretly, by way of Jamaica, exactly like an article of contraband. This situation continued until 1741, which might be considered the beginning of platinum's history, for from that time on it gradually emerged from its mystery and ambiguity to attain its rightful status: as the rarest, the most inalterable, the most precious of metals.

In his memoirs, Giovanni Giacomo Casanova gives a detailed account of the attempt he made, with the marquise of Urfé, to turn platinum into gold.

THE GARLAND STYLE
From Its Origins to 1914

Numbers possess a magical and mysterious power all their own. One example is the way every new century immediately distinguishes itself from the last by turning away from it, as if it abruptly changed its point of reference on hearing a clock chime the hour. Examining this phenomenon more closely, it becomes apparent that this turnabout has its origins in far-off events and trends. However, the innovations that introduce a new era often seem to come out of nowhere. The French Revolution turned the Age of Enlightenment toward Romanticism and idealism with eleven years to spare, but then Napoleon set the course of the nineteenth century in a totally unexpected direction.

At the turn of the century, when Cartier moved his boutique in 1899 to Paris's most elegant street, the rue de la Paix, France, Europe, and the rest of the Western world had enjoyed thirty years of peace. The century that would see two horrendous world wars was born amid ominous or, at best, contradictory signs, for those who could read them. In France alone, the final years of the preceding century brought two foreboding events: in 1898, the Fashoda incident, a serious diplomatic crisis symptomatic of the colonial mentality, in which relations between France and Great Britain hovered on the brink of collapse; and in 1899, the culmination of the Dreyfus affair, which had torn apart the nation's conscience for five years, dividing it into two sides that confronted each other with the violence of ideological civil war. These two emblematic events cast menacing shadows on the new century, bringing to the surface at least three symptoms whose development would prove to have fatal consequences: nationalism, anti-Semitism, and the crisis of colonialism and, therefore, of Europe.

After many years of peace and economic growth, society was in enormous flux. Rampant industrialism brought forth its monsters — urban-slums, miserable living conditions for the masses, the lack of regulated working conditions (initial palliatives were timid and small scale; for example, the Millerand Act in France, which set the work-day to a maximum of ten hours, passed in 1900, confirming the symbolic power of numbers) — but it also produced miracles. The middle class was at its height, and the aristocracy had many years yet before World War I signaled the end of its glory. The glamour of the blue blood and the magic of money irresistibly attracted one to the other, exemplified in the union of Boni de Castellane (an aristocrat without means) and Miss Gould (an American heiress, as plain as she was wealthy). To a friend who criticized her physical attributes, her husband said of her, "In the light of her dowry, she doesn't look bad at all."

Those years saw the spread of snobbery, the serendipitous union of intelligence and frivolity, which found its most refined and ironic bard, not accidentally, in the Parisian Marcel Proust. Proust himself was an exceptional representative of snobbism, and his character Swann its most perfect literary incarnation. The so-called species, an extraordinary hybrid of assets (money, culture, artistic or critical talent, conversational art) and deficit (one only: the lack of a title; the word *snob* is a contraction of the Latin expression *sine nobilitate*), suffered an imbalance among various natural gifts that had meaning only in a particularly sophisticated society that was also, of course, affected by a profound imbalance.

The Belle Époque was a period of great contradictions and contrasts not limited to the social stratum. In the preceding century, science had made prodigious progress. An increasingly advanced technology invaded every aspect of daily life and made possible not only items of luxury but also commodities previously beyond the reach of kings. Accompanying this development was a positive attitude, strongly steeped in the faith in progress, skepticism toward religion, and a taste for the concrete. The individual of the second half of the nineteenth century breathed an atmosphere of certitudes, of measuring instruments, of figures and formulas. And the new century presented itself as representing the triumph of reason over superstition, of science over religion, and of humanity over nature.

The loftiest and most fascinating symbol of this faith in progress, technology, and science is the Eiffel Tower, built in 1889 for the Paris World's Fair and never destroyed (as most other such special-occasion buildings were). It looked like an architectural forerunner of the future, very high and very light, stretching to the sky as to the future. A challenge to God, it was said, but to art as well: its engineering made an aesthetic statement, proving, for the first time, that technology could create a thing of beauty.

On the threshold of the twentieth century, this attitude came into question. A subterranean movement began to gnaw away at the reassuring tenets of positivism. While barely perceptible to the average person, it became noticeable in the posh and apparently carefree social circles of Paris and London, in the person, for example, of the artists that frequented them. Psychoanalysis was on its way, the scandalous champion of the uncon-

Designs for the central peak of a tiara and several brooches in platinum and diamonds. Archives Cartier Paris, ca. 1900

scious, which is to say, of an indefinable essence, difficult to pin down, impossible to capture with scientific instruments. And by now society's unconscious was indeed agitated.

Despite the stability that accompanied the long period of peace and the expansion of well-being and wealth, in the artistic domain the society of the Belle Époque experienced an extraordinary period of renewal, of the breaking of rules, of fertile disorder. Within a few years, all the arts seemed to question their own fundamental principles: figurative art rejected the figure, music revolutionized the idea of harmony, poetry did away with meter, and the novel unraveled its weaving and knotted unaccustomed threads, abandoning the surface to explore the depths. In the same circles, enthusiasm for newness cohabited with deep roots in tradition.

Cartier was an innovator by nature, preferring to start rather than follow trends, but he was above all a jeweler, in those years on his way to becoming the world's most famous. He served the elite of that period's society, those who wore the most sophisticated outfits, kept the most refined

Opposite: Edward VII of England with Queen Alexandra. During one of his frequent visits to 13, rue de la Paix, Edward VII dubbed Cartier "the king of jewelers and the jeweler of kings."

company, frequented the most exclusive haunts and the most luxurious establishments. Cartier the innovator could not lose sight of the classic; on the contrary, he had to present a transformation of taste that fulfilled and flattered it. The Garland style, which seemed to represent a return to the past, employed the radically new material platinum, thereby striking a balance between conservatism and innovation. Cartier seemed to breathe the air of his time and to metabolize it through his own genius: he inhaled contradictions and exhaled harmonies.

In the area of the decorative arts, the final decade of the nineteenth century was dominated by Art Nouveau, a fertile concentrate of contradictions, which, by the same token, perfectly represented the period from which it emerged. In the first place, this cultural movement, so cosmopolitan that it invaded all of Europe simultaneously, leaving important traces of its passage everywhere, had no single international name. It had as many names as the countries in which it flourished: in France, Art Nouveau; in Italy, *stile floreale* or *stile Liberty*; in Great Britain, the Modern Style; in Germany, Jugenstil; in Austria, the Secessionists; in Belgium, the *Velde stile*; in Spain, *modernismo*. Therein lay its first contradiction.

Art Nouveau arose in reaction to the eclecticism that had dominated architecture and all the applied arts during the whole of the nineteenth century. If any century lacked a distinguishing style, it is this one, which learned everything, recast it all, and used it all up. The modernists, rebelling against the prevailing historicism and the extensive taste for pastiche, proposed a graphic and ornamental style that drew its inspiration directly from nature—not the trite scholastic nature of academic decoration but a nature treated artistically in a dreamlike manner, stopping short of hallucination or nightmare. A nature in movement, in transformation. A nature, whether animal or vegetable, that twists and curves in a spiral. The word that characterizes Art Nouveau is *twining*. Its lines never go straight, but invariably curve, making tracing arches, loops, scrolls. Nature is actually suggested rather than represented—intertwinings of leaves and branches, swellings of flowers, snakelike lightning bolts that invade every corner of the design. And therein lies another contradiction: the naturalism of the subject contrasted with the extreme artificiality of the composition.

Art Nouveau grew out of an earlier movement that preceded it by several decades, the Arts and Crafts movement. The latter, associated with a specific nationality and origin, was born in England, where it was joined by William Morris. Following the example of the members of this group, modernists revolted against the vulgarity of objects made in mass production by machines, championing instead a return to the handmade object using the techniques of fine craftsmanship. They had as their goal the rehabilitation of furniture, which they hoped would bring about the greater rehabilitation of the domestic space and daily life. Paradoxically for this reason, detractors considered modernists enemies of modernity, of progress and industrial development. More paradoxically still, one of the prime characteristics of Art Nouveau was its interest not only in new techniques but also in new materials, such as iron, glass, cement, the first synthetic resins—the standard materials of industrial production. The supposed modernist enemies of industry supported a strong alliance between industry and art, striving for a means that would make beauty available to all. The primacy of design found in Art Nouveau its cradle and its crucible.

Cartier made jewels in the Art Nouveau style only sporadically, but it would be misleading to say that the style did not interest him. The problem was that with its complex design, its

sinuous lines, its tendency toward volume, it did not lend itself to the technical aspects of jewelry production (in fact, most of the jewels realized by modernist artisans were made for exhibition and were practically impossible to wear). Cartier created jewelry for a thin stratum of society that, during the Belle Époque, were worldly and privileged individuals with sumptuous living standards, including a steady diet of balls and festivities. The Art Nouveau style was not adapted to their jewels for two basic reasons, one of a socio-aesthetic nature, the other of an aesthetic-functional nature.

To begin with, the elite that corresponded to Cartier's clientele was more interested in classicism than in modernism. While flirting with modernity in passing associations with artists (the elite liked to surround itself with artists, living emblems of modernity), they feared in modernity, more or less consciously, a new broom that would sweep them away. This explains their adherence to the past, and aesthetic of elegance and lines that, while not entirely bereft of the flowery embellishments of the preceding century, tended toward purity rather than elaboration. In addition, this group wanted to show in the slightest gestures, words, objects, or customs the almost legendary rarity of its privileges. And since of all the outward effects of wealth the jewel is the richest, the search for the sublime, the expensive, and the precious becomes even more selective. The great ladies of the Belle Époque wore evening jewels that were actually substantial inheritances transformed into diamonds.

For Cartier, the aesthetic-functional reason took precedence over the socio-aesthetic one. The diamond dictated the jewel, determining the way it should be conceived in order to showcase its fire completely. This called for a clear, distinct design without unnecessary elements. In this light, Art Nouveau was unsatisfactory to Cartier, given its indefinite lines and twists. For him, the jewel required a perfect balance between filled portions

and empty spaces, and diamonds in particular required space in which their brilliance could expand. In a jewel, a design that is too complicated both dulls the brilliance of the stones and is itself eclipsed by that brilliance.

Louis Cartier, who took the reins of the business into his young hands in 1898, was a man of culture and refinement, highly interested in new ideas and gifted with a keen business sense as well. His first stroke of genius, which brought the Maison Cartier enduring recognition, was to relaunch the Garland style.

The Garland style (its name may come from the Renaissance painter Domenico Ghirlandaio, whose father was a Florentine goldsmith) derives from the Louis XVI style and was not really a novelty in jewelry. Its success began in the Second Empire, when Empress Eugénie had new settings made for all of the jewels of the reigning house in the style of Marie-Antoinette.

At that time, it drew its inspiration from the classical and rigorous style of eighteenth-century France. However, by the second half of the nineteenth century, this restrained style had been mindlessly copied and reworked to the point of falling into overwrought mannerism.

Cartier went back to the source, to the purity of design that is the keynote of the authentic eighteenth-century French style, which he loved above all others. The Garland style as relaunched by Cartier at the beginning of the twentieth century, which would remain popular until World War I, was a light, purified style. Its design was clear, its lines distinct, taut, without flourishes. Its curves were always elegant, soft, never tortuous. Empty spaces set off filled ones, allowing the stones room to breathe and lending volume to their sparkle.

The problem with "white" jewels for evening was the metal used to support the stones. Silver was traditionally employed for this purpose, using gold on the side that touched the skin and silver for the mount. However, silver presented two serious defects: its excessive malleability, which entailed heavy and complicated settings, and the tarnishing caused by rapid oxidation. Combined, these defects produced a heaviness and darkness that tamped the fire of the diamonds, dimming their splendor and rendering them less iridescent.

Cartier's genius lay in speculating, with great boldness and not the slightest hesitation, on the new metal platinum, which jewelers had been leery of using, as the nature of this precious metal was still barely known. Fortuitously, platinum suited the jewelers' needs perfectly: strong and highly resistant, it could be used in small quantities, and

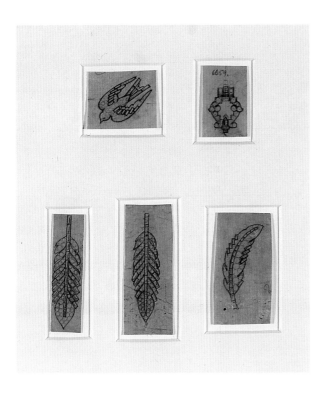

Designs for five brooches in platinum and diamonds. Archives Cartier Paris, ca. 1914

near-invisible as a setting to support the stones. Nor was it susceptible to oxidation—far from diminishing the fire of diamonds, it enhanced them without calling attention to itself.

It can thus be seen how Cartier, while apparently remaining indifferent to the spread of Art

Nouveau, had actually taken hold of it and learned its most radical and lasting lessons—form and substance.

From a formal point of view, the experience of Art Nouveau taught a valuable principle: that the repetition of worn-out, historicizing, and quasi-architectural ornamental motifs from the past had been surpassed by a new style, whose power resided in its clean lines and clear, strong design. While Cartier rejected the Art Nouveau line, with its most twisted configurations, because it was incompatible with the requirements of jewelry, he accepted the uninterrupted flow of its strong curve.

The other lesson that Cartier drew from Art Nouveau was to consider new materials. The modernists had demonstrated that industrial materials such as iron and cement were fully capable of creating beautiful effects. Among the noble metals, Cartier confidently singled out platinum. The parallel between the former materials and the latter would later prove to be more pertinent than he might have suspected at first view, considering that platinum was then, at the dawn of the new century, mostly used in industry and scientific laboratories. After a century of intense experimentation with the new metal, its characteristics—a high melting point, resistance to acids, and catalytic properties—made it invaluable in a number of applications. The large utensils in which sulfuric acid was distilled, special instruments that measured very high temperatures, the filaments of the earliest incandescent light bulbs, the international standards of metric weights and measures—all were made of platinum. At that time, the word *platinum* evoked science rather than jewelry, technical progress rather than style, function rather than beauty. In fact, it was platinum's association with industry that caused jewelers of the time to look on it dubiously. It was not until 1912 that France officially named it a precious metal and gave it a hallmark, in the form of a dog's head.

Louis Cartier's decision to combine diamonds and platinum for his Garland style jewelry was therefore a courageous act that went against his clientele's initial mistrust of platinum. This mistrust, however, was easily overcome by the metal's beauty and luminosity, as well as the extraordinary elegance of the jewels that the company began to produce without any hesitation. In Cartier's hands, platinum became synonymous with lightness.

Before wedding platinum and diamonds so successfully in the Garland style, Cartier had made occasional use of platinum, as had other jewelers, including such distinguished names as Fabergé, Bucheron, Mellerio, Tiffany, and Van Cleef & Arpels. Its first mention in the archives of the Maison de Cartier dates back to 1853, when Cartier used it to make a series of shirt buttons. In the following decades, its use remained irregular, as is normal with a material still considered experimental. It found its way, combined with gold, into cuff links and tiepins, some rings, and a watch in the form of a heart—that is, for accessories rather than jewels, not particularly important objects. Even after platinum was named a precious metal, it did not evoke great enthusiasm, and, although technically unnecessary, it continued to be paired with gold for the most part. This also confirms that when used alone platinum was not yet ready to confer on an object of daily use the glamour or precious character of a small jewel.

Cartier's daring elevation of platinum at the very beginning of the twentieth century was unquestionably facilitated by its technical evolution. For at least a century, platinum had aroused the curiosity of scientists, particularly for its extremely interesting physical properties. But it was these same properties that made it very difficult as well as costly to obtain in pure form, which is the only way it could be used in jewelry for important creations. Its exceptional resistance made it very hard to separate from the other metals of the same family—iridium, osmium, rhodium, ruthenium,

and palladium—with analogous properties and always found combined with it in its natural state. It was not until the very end of the nineteenth century that a German physicist named Heraeus discovered how to carry out the process of refining platinum in order to obtain a fairly pure form of the material. Only then could it be found on the market in sufficient quantities to satisfy jewelers' needs. Significantly, Louis Cartier pounced on the newly available pure form, indicating the interest he had taken in the material long before that point and his awareness of the advantages the metal had to offer.

During the fifteen years between the opening of the shop on the rue de la Paix and the beginning of the war, Cartier made a great many jewels in platinum and diamonds, some with large natural pearls of exceptional sheen. They were a triumph of white, of light, of lightness. They also signaled the triumph of Cartier itself: already one of the favorite jewelers of Parisian high society, he was on his way to becoming the most famous jeweler in the world. This series of events holds more importance than meets the eye.

In 1902, Edward VII, "eternal Prince of Wales," finally became king. Through the years, he had patronized Cartier and was a great connoisseur of its creations. For the coronation, the French jeweler was besieged by the great ladies of the aristocracy looking for new settings for their sumptuous parures. That same year, Cartier opened a London branch. In 1904 he was named official purveyor to the English court and created for Queen Alexandra a superb *résille* necklace made of diamonds and platinum, with a laurel leaf motif, which was immortalized in a court portrait by the well-known painter François Flameng. In 1909, Cartier established a New York branch.

Of course, the success of what might be called "Operation Garland" was tied not only to the daring combination of diamonds and platinum but also to another association at which Cartier had always excelled: between jewelry and clothing. The rue de la Paix, where the jeweler had moved the shop, was the street of haute couture. One of the two sons of the great couturier Charles Worth, the undisputed leader of the period's stylists, was Louis Cartier's father-in-law (the other became the father-in-law of Louis's sister Suzanne). Fundamentally, Cartier's interest in women and in changes of styles, as well as of life-styles, was unflagging as well as subtle.

The woman of the Belle Époque always covered her body completely, but not her shape. She had given up crinolines, preferring to swathe herself in soft and supple fabrics that showed off her curves. It was said at the time that the curve was the great discovery of the moment. "Twining" was not restricted to Art Nouveau; it applied equally to ladies' fashion. Garments fit the body like a glove, creating silhouettes that traced alluring sinuosities in space. The corset compressed the diaphragm and pushed up the breasts. The small waist was like the center of an elongated S, midway between the generous curves of the breasts and the hips. Here, as in the purest expressions of ornamental design, found in Art Nouveau and in the Garland style, the line is firm and strong. Eroticism combined with modesty to cover these desirable round forms with ethereal layers: trains, large hats with veils, close-fitting gloves, transparent laces, all flowing and graceful.

The excitement resided entirely in the contrast between what could be guessed, deduced, or imagined and what could not be seen at all—that is, in the play of light and shadow. And nothing could be brought to light more brilliantly than the jewels in diamonds and platinum by Cartier, which were in themselves pure light.

Set in platinum, the diamonds described strong and luminous curves, apparently suspended in midair. Tiaras (such as those reproduced on pages 80, 92, 94–95, 98, and 99), necklaces (the

Design for a pendant in platinum and diamonds.
Archives Cartier Paris, ca. 1912

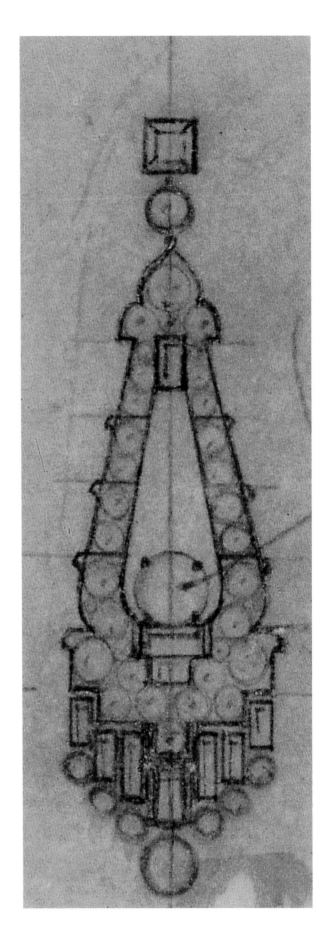

famous one belonging to the Spanish dancer La Belle Otero, on pages 78–79, alone suffices to represent all), brooches (pages 82, 83, 88, 89, 96, and 97), and earrings adorned a woman and her clothes. Skin, displayed in those very restricted areas allowed by convention—the face, the décolletage, the hands (sometimes)—exerted attraction, enhanced by the brilliant jewelry.

It should be emphasized that without platinum Cartier could not have achieved the airy lightness of the Garland style jewels, with their lacelike evocation. Resistant, easy to work, and requiring modest amounts, this metal offered superb results. In addition, its silver-white sheen, which could be called neutral, enhanced the white fire of diamonds without dulling its refractive qualities. Cartier employed a fine-quality platinum, the purest and whitest, and turned it into settings made of fine wires or of tiny shagreen rings (the *millegrain* setting) that broke up the metallic reflections of light into hundreds of quivering glints. These two characteristic elements, lightness and the splintering of the white light, contributed to the success of the enterprise: to "serve" the diamond, to support it and emphasize it while leaving it alone in the spotlight, as if it were the stones themselves, held down only by gravity or magic, that delineated the sinuous design of the jewel.

Seeing certain necklaces *en résille*, which can cling to the skin, covering the neck and the triangle of décolletage as with a fine network of silk, it is hard to believe that the setting of the jewel is made of a rigid, heavy metal. All that is visible is a sparkling textile that skips from one stone to the next, as if the diamonds possessed the ability to spin flashes of light and to attach themselves to the web like spiders.

Edward VII, successor to the "immortal" Queen Victoria, acceded to the throne at the age of sixty. His reign lasted less than a decade. If his coronation marked the aesthetic and social triumph of the Garland style, that of his successor,

George V, in 1911, definitively sanctioned it. Two months before the ceremony, in April of that year, a memorable exhibition took place in the rooms of Cartier London, on Bond Street. Nineteen tiaras by Cartier were put on display, all of them destined to shine on the foreheads of as many representatives of the fine flower of European aristocracy at the coronation. The exhibition, which charged an entrance fee (for charity), attracted thousands of visitors. From one coronation to the next of these two kings of England, the union of platinum, diamonds, and Cartier seemed to describe, by means of interwoven brilliants, the strengthening of the Entente Cordiale between London and Paris.

Beneath the glitter of festivities and tiaras, the war was on the watch, and during all that time, Cartier was weaving together another revolution in style.

Design for two brooches in platinum and diamonds. Archives Cartier Paris, ca. 1907

Platinum sugar bowl by Marc Étienne Janety, goldsmith to the court of Louis XVI. Collection The Metropolitan Museum of Art, New York

*A*ntonio de Ulloa, a Spanish officer who had returned from a geographic expedition in Ecuador, in 1748 published an account of his journey. It was the first in European history to talk about platinum from the vantage point of direct observation. In the same period, the interest of scientists in the new metal, aroused by the few samples that had become available, expanded in many parts of Europe.

English scientists in particular scrutinized this as yet unknown metal. Charles Wood and William Brownrigg carried out the first scientific experiments in the laboratory, whose results were reported to the prestigious Royal Society by a third scientist, William Watson. William Lewis was the first to describe, in his publications, the physicochemical properties of platinum with scrupulous scientific precision. It was left to the Swede Teophil Scheffer to propose, in 1751, that platinum was not, as suspected up to then, an alloy of gold and iron but an entirely separate metal. This hypothesis was rejected to the end of the century, even by such celebrated scientists as the French naturalist Georges-Louis Leclerc de Buffon. Scheffer was also the first to melt platinum, by adding to it small quantities of arsenic.

More and more was learned about platinum. The German Andreas Sigismund Marggraf succeeded in separating it from the other metals of its group, with which it always occurred in its natural state, obtaining a relatively pure form of the metal. The Frenchmen Macquer and Antoine Baumé, the first to realize platinum's malleability, managed to beat it into sheets. Karl von Sickingen produced platinum wire in 1780; he could thus be said to be the first to understand and exploit its extraordinary ductility. For the moment, all these advances were confined to laboratories carrying out scientific research and experimentation. Up until the end of the eighteenth century, only minimal amounts of platinum were dissolved, refined, and worked. Any possibility of commercially exploiting the product remained far in the future.

One of the primary figures in the history of platinum was the French goldsmith Marc Étienne Janety. His work stirred interest in platinum among jewelers for the first time, and it began to take on the aspect of a precious metal. In order to melt it, Janety used the only method then known, introduced by Scheffer, of carefully adding arsenic. The French artisan, goldsmith to the Crown before the revolution, perfected that method and obtained a particularly malleable form of platinum that he used to create precious objects, such as a sugar bowl, today owned by the Metropolitan Museum of Art, New York, and a coffeepot, which has been lost. Later, in 1799, the republican government entrusted Janety with the task of making the official standards of weights and measures in the metric system.

Up to about 1825, when the Russian mines of the Urals began to be exploited, all the world's platinum came from the gold-bearing sand of Colombia, which then belonged to Spain. King Charles III and Charles IV apparently became entranced by the new metal's mysterious charms. Charles III gave Pope Pius VI the first sizable decorative object in platinum of which mention can be found—a chalice almost a foot high and weighing over four pounds, today housed in the Vatican Museums. Charles IV, in his turn, had a Platinum Room made in the royal palace at Aranjuez to contrast with the Gold Rooms, which then lost much of their originality in comparison.

Of all the monarchs, it was Marie-Louise of Austria, second wife of Napoleon, who most clearly forecast—a century in advance—the fashion for jewelry in platinum. The empress's sumptuous ceremonial gowns, in the purest Empire style, were embellished with very fine platinum threads. They can be seen in some official portraits, especially one of the gowns Marie-Louise wore when she was awarded the Italian duchies of Parma, Piacenza, and Guastalla at the Congress of Vienna. One of these fabulous outfits can still be admired at Parma's Museo Glauco Lombardi: a white silk dress covered with tulle, scattered with platinum threads and decorated on its borders with vines and cornucopias executed in platinum threads. From the waist flows a long silk train, embroidered with the same motifs in the same material. Still far from being recognized as the king of metals, platinum was already on its way to becoming the metal of kings.

Portrait of Marie-Louise of Austria, wife of Napoleon, the French emperor, and mother of the king of Rome. The empress had several dresses embroidered with platinum thread. Collection Museo Glauco Lombardi-Parma

ART DECO
1915–1925

At first sight, the two brooches illustrated on page 81 might seem out of synch with Art Deco. Made between 1906 and 1907 by Cartier, these geometric creations in diamonds on platinum are in full Garland style, yet the curved line looks as if it were suddenly straightened out.

Such pieces further an understanding of the depth of the roots where the tree of a cultural, artistic, and intellectual movement as complex and rich as Art Deco draws its impulse and vitality. The "straightening of the curve" was the first step toward the affirmation of a new style. Another was the explosion of color. However, from a strictly graphic point of view (and Art Deco is a style in which that aspect dominates), the transformation of curved lines into straight lines, of concavities into angles, and convexities into corners already constituted a decisive step.

If it is true that every period contains the fruits of the preceding one and the seeds of the one to follow, it is necessary to take a closer look at the Garland style to discover the signs of an apparently paradoxical future: from sinuous to angular. Actually, this transformation can be seen in two pronounced characteristics or, at least, tendencies of the Garland style—an inclination toward abstraction and a respect for symmetry.

This inclination toward abstraction, still relative, can be better appreciated after one more comparison between the Garland style and Art Nouveau. Both styles drew their inspiration from nature as model and drew away from nature in interpretation, but in two directions that were in a certain respect opposite. Art Nouveau wanted to be more natural than nature itself: this was an instinctive art form, one that rebelled against orthogonals. It adds more than it subtracts. The Garland style, on the other hand, clearly prefers subtraction to addition, and order to disorder. In it, nature often becomes unrecognizable through pruning, not through an excessive luxuriance. Of the flower, all that remains is its grace, its seduction. In this sense, one can already talk about abstraction in respect to the Garland style: of the physical model, all that is left is a refined graphic projection.

Even more evident is the respect for symmetry in the Garland style. Flowers and knots, bluebells and garlands, tassels and laurel leaves unfailingly fill to the brim the two halves on either side of the central axis, mirror images, their numbers equally divided. The concern for balance extends to the way the gems are distributed— according to their weight, that is, the number of carats; pearls, according to the number of grains; and diamonds, according to their "magnitude," that is, the intensity of their fire.

Symmetry represents, even more than does abstraction, a distancing from the model of nature. In nature, symmetry is an ideal. Facing this more or less noticeable gap, art, as imitator of nature, can choose between two divergent paths: that of the concrete, which tends toward disequilibrium, and that of the abstract, which with its impulse toward the ideal divides all the volumes into two perfectly equal halves.

In conception, the Garland style falls between the extreme refinement of organic traits and the extreme idealization of forms. Much less concerned with the past than it appears to be at first glance, it works by means of its soft and gracious curves toward the ultimate fate of all styliza-

tion: geometric form.

A comparison between the two brooches cited above, and the three brooches shown on page 125 (creations of the years 1923–25) will illuminate some especially interesting aspects of an evolution that took almost twenty years and continued up to the beginning of World War II. At this point, it was focused on issues of colors and volumes rather than forms and design.

The 1906–7 brooches are basically geometric forms, respectively, a hexagon and a lozenge. Yet, within this "frame" delimited by straight lines and abrupt turnings without curves, the flourish of the Garland style is still present in the form of its attenuated symbols. The attenuation is such that they cannot even be named, much less classified, except for the laurel leaves arranged along the outer edge of the hexagonal brooch, which introduces a lively wave into the rectilinear thrust of each of its sides. Otherwise, nature is evoked by simple conventional references: rays of sunlight, small spheres resembling berries, fine arches like the corollas of hypothetical flowers. It might be described as nothing more than a Garland style that has taken the bold step of enclosing itself in a geometric space.

By contrast, the three 1923–25 brooches, created more than fifteen years later, might appear distinctly less angular, but the spirit of these objects are actually much more geometric. At this point, the abstraction has been sufficiently mastered that the line can again curve in on itself without losing its graphic clarity, its implacable Euclidean determination. Nature, in terms of a living repertory of images, is so distant that where its apparition emerges, the evocation seems to take on a lightly ironic tone. Thus, for example, in the brooch on the bottom of the page, the central ring in rock crystal that forms a broken ellipse (called *en tonneau*, or barrel-shaped), carries at its ends

Italian opera singer Lina Cavalieri was one of Cartier's most enthusiastic customers.

two motifs in diamonds that represent the head and the tail of a dragon. But the sinuous curves of the reptile have become almost right angles, a play on broken lines. The nature that the object imitates has retreated to become as far distant as a lost galaxy; here, it is nature that seems to imitate geometry. This dragon does not spout fire, it

Designs for several necklaces and pendants in platinum and diamonds. Archives Cartier Paris, late 1910s

hibernates within its arthritic skeleton, its stiff, orthogonal articulations.

Platinum was made to support this abstract and rigorous jewelry. Easy to work yet strong, ductile yet resistant, it adapted perfectly to all the geometric shapes and lent its lightness as well as its concrete nature to all the abstractions. This time it was not restricted to diamonds, as in the Garland style. From the beginning, Art Deco used color, and platinum's superb neutrality, its capacity to reinforce the brilliance of gems and the most varied materials without interfering with their reflections, nuances, or chromatic effects, came to the fore.

Going by color scheme, a first type of inspiration can easily be singled out: creations in black and white. The white, of course, consisted of platinum and diamonds, as well as rock crystal, a less noble material but attractive for its transparence and luminosity. Cartier used it freely in the 1910s and 1920s, and it contributed significantly to the extraordinary success of the famous "mystery clocks" (two of the three brooches on page 125 mentioned above furnish examples of rock crystal included in black and white jewelry). For the black, primarily two materials that, in a sense, could be considered new were employed—black enamel and onyx (see the brooches on page 125 and the bracelet watch on page 136). An interesting if morbid note: the vogue for black and white jewelry began in 1911, when the sinking of the *Titanic* brought about a rebirth of the Victorian style of "mourning jewelry" among New York's high society, which saw it as a way of commemorating those lost at sea while keeping in luxurious style.

Another kind of inspiration emerged from the Cartier workshops of the period: colored jewels or gems combined with the white of platinum and diamonds and, eventually, the black of onyx. Emeralds (pages 122, 123, and 139), sapphires (page 128), turquoise (page 118), as well as jade (page 119), coral, mother-of-pearl, and combinations of emeralds and sapphires (pages 120–21) or of rubies and emeralds (the parakeet on page 129 and the Egyptian brooch on page 133)—all seemed to announce the explosion of color in the later "tutti frutti" creations.

Art Deco received its name very late, from the 1925 exposition in Paris devoted to "Arts Décoratifs et Industriels Modernes" (Modern Decorative and Industrial Arts). At that time, it had already reached its full maturity; it could even be argued that from a graphic point of view it would see no subsequent development. In every

field—architecture, furniture, graphic art, household arts, fashion, and industrial design—Art Deco, after a period of incubation that stretched from about 1907 or 1908 to the beginning of World War I, realized its triumph as a style and as a fashion immediately following the war and throughout the 1920s. In all its expressions, it was characterized by geometric shapes, strongly marked lines, the clean carving of surfaces, by the openly "decorative" and "abstract" elegance of its composition. (In fact, the Russian painter Sonia Delaunay Terk named the outfits she launched in 1916 "abstraits.")

In the previous chapter, the Eiffel Tower was named as the most significant example of the faith in technical progress with which Europe prepared itself for the new century. New York's Chrysler Building is certainly appropriate to symbolize a period that was, in its way, indecipherable, cut in half by the worldwide conflict and corresponding, on the one hand, to the opulent final years of the Belle Époque and, on the other, to the Roaring Twenties that followed the war. The Chrysler Building is one of the higher skyscrapers in Manhattan (at the time it was built, in 1928, it was the highest), so it shares with the Eiffel Tower a clear vertical dominance, an elegant and lively structure. But the technical beauty of the Eiffel Tower, which managed to transform iron, the symbol of industrial technology, into a material of not only solidity but also, by means of the airy lightness of its framework, of aesthetic importance, was only a memory. The beauty of the Chrysler Building arose as much from the Art Deco ornamentation of its skin as from the audacity of its conception: humanity continued to climb to the clouds, but it no longer believed that the industrial society would bring well-being to all and that beauty resided in the simplest of objects—which

Victoria Eugénie, queen of Spain, is wearing a diamond and platinum tiara with seven pearls made by Cartier in 1920.

the Russian constructivist architects, in their isolation, still maintained at the time. Beauty was the heritage of some. Art Deco, in a certain sense, was the definitive reaction of the comfortable bourgeoisie to the upsets of the first two decades of the century, which found their artistic and cultural expression in such movements as Dada and Futurism.

All the protagonists of Art Deco, from the couturier Paul Poiret to Sonia Delaunay, from the illustrator Erté to the artisan René Lalique and, naturally, Louis Cartier, were completely aware of the cultural climate in which this art developed, as well as the conditions that promoted its spread.

The year 1907, to which can be traced the earliest signs of the future Art Deco, is also the more or less official date of birth of a movement that would take on a singular importance in the history of Western art and deeply influence all subsequent formal explorations: Cubism. This movement, led by Pablo Picasso and Georges Braque and with such eminent adherents as Fernand Léger, Robert Delaunay (husband of Sonia), and Juan Gris, arose out of the urgent need to rethink pictorial space. The Cubists rejected the space according to traditional perspective and went in quest of a three-dimensionality that, instead of residing within the space of the representation, could be centered on the object's representation in volume.

The destruction of perspective, which was implicit in such earlier movements as Impressionism and Fauvism, clearly demonstrates the profound blows that the intense artistic experimentation of the early twentieth century dealt the scientific and positivist beliefs of the second half of the preceding century.

In their attempt to offer a "simultaneous" vision of the object represented, that is, a global representation that no longer stretched invisible lines between the observer and the plane of projection, the Cubists, with Braque and Picasso at their

Designs for a flat brooch in platinum, diamonds, and onyx and two rosette brooches in platinum and diamonds. Archives Cartier Paris, ca. 1915

head, took the object apart, separating it into all its constituent pieces. They opened all the surfaces to penetrate the object's volume and scattered on the canvas the broken planes, cut into facets, that resulted from this "analytical" operation. The dismembered object thus opened itself to the mystery of the space that it contained.

In the apparent disorder that invaded the pictorial space of the Cubists was in reality a need for a rational order of a higher sort, capable of liberating art and freeing humans from their exclusive relationship with the givens of reality perceptible through their senses. Precursor to abstract art, Cubism affirmed the primacy of geometry as an instrument of perception as much as of expression.

At the same time, machines were penetrating the human world and revolutionizing daily life. While faith in progress was no longer absolute, technology did not cease to transform society. Even the war, with its unparalleled destruction and millions of dead, pushed forward technological progress.

It was certainly no coincidence that Louis Cartier took the horizontal section of the tank as his inspiration for the prototype of his famous Tank watch in 1917. The essentially functional character of the machines of war possessed a naked, unvarnished beauty all its own. And in this specific application, more than before, could be seen the beginnings of the geometric form. Cartier found the watch seductive, particularly the bracelet watch or wristwatch, precisely because of the way it naturally yielded to the laws of geometry. In this case, the cleanness of the design stemmed not simply from a formal choice but also from a functional necessity, as well as the requirement that the watch be easy to read and easy to wear. Right after the war, the production of wristwatches greatly increased, and platinum played a major role, as confirmed by the objects illustrated on pages 114–15. Among them can be seen some of the first jewels (what might be called functional jewels

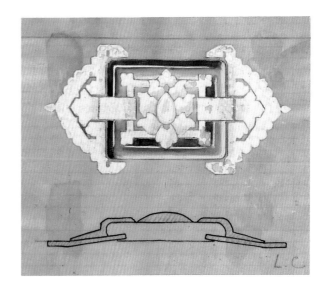

Design for a brooch in platinum, rock crystal, and diamonds. Archives Cartier Paris, 1925

rather than watches) that employ platinum all by itself, no longer as the accompaniment of diamonds or gems, indicating that it was finally recognized as a precious metal. In its solo form, platinum proved singularly effective at enhancing the object, emphasizing its clarified geometric lines. With its clear metallic identity, it seemed to bring out, better than gold, the modernity that the wristwatch both promoted and incarnated.

The war brought a geometry of a different kind, more indirect but more explosive. With the absence of men who had gone to the front, women throughout the Western world had the opportunity to hold new positions involving responsibilities, work, and independent activity which gave them a new awareness and the hunger for their long-denied freedom. By the end of the war, women had no intention of being shut up again in the cage of traditional feminine attributes. Combining with the euphoria that overtook Europe on achieving what Pope Benedict XV called the "futile massacre," the emancipation of women provoked an uproar in the social and cultural life of the 1920s.

The new woman, active, dynamic, independent, who worked, played sports, and drove a car, found her pre-war clothes completely outdated.

Skirts grew shorter, stopping just below the knee, and legs were revealed for the first time, not so much to show them off or to unleash their power of seduction as to continue the transformation to the new tubular silhouette. Legs not only symbolized movement, which was the keynote of the period, they were living geometry. The straight line, so powerful at the time, was further manifest in legs, barely softened by the attenuated curve of the calf. From all evidence, other lines and curves did not reveal themselves but tended to remain hidden, which confirms that skirts were shortened for reasons connected not with seduction but with geometry. So while the style of the Belle Époque had discovered the curve, that of the Roaring Twenties rediscovered the straight line and asserted it with a newly found ease. "The straight line is a mode of expression" is how Coco Chanel, grand priestess of fashion in the 1920s, put it—a highly effective mode of expression, especially representative of the woman who wanted to follow the direct path.

The woman of those years in France was given the funny and explicit nickname *garçonne*, or girl-boy; in America she was known as a gamine. This was a woman masculinized, but in a youthful way, that is, with her roundness removed and transformed into a long-lined silhouette. The tubular cut of her clothes highlighted angles rather than curves and went for the gesture, the form in movement, rather than the grace in the beauty of forms: no longer "stay," but always "go." Yes, the straight line was certainly a mode of expression, and the hair cut short (*à la garçonne*, in fact) made that expression even more explicit and pronounced. What it expressed was, evidently, the equality of the sexes.

Cartier had developed a special relationship with women, which led him to emphasize the wearability of his jewels. He knew that the jewel was a clothing accessory, which could be said to be an accessory of the body—and this, paradoxically, is what differentiated him from the other jewelers of his time and what made him the greatest. Cartier himself supplied the most eloquent demonstration of this premise, at the famous Paris World's Fair of 1925 (the one that gave the already mature style Art Deco its name). He decided to exhibit his creations not in the Grand Palais, with the other jewelers, but in the Pavillon de l'Élégance, with the couturiers. This gesture could be seen as a polemical maneuver, but it was the clear statement of a principle: the jewel should not stand out from the clothing, nor, even more important, from the person wearing it; rather, it should harmonize with one as well as with the other, enhancing skin color as well as the fabrics it accompanied.

Proceeding from this principle, Cartier, who had been the first to attempt the "rectification of the curve," now tended to play down rather than to accent the angular nature of Art Deco's line. Certainly, he seemed to be saying, the straight line is a mode of expression, but the curve, if it is firmly geometric, can alternate with the straight line, can accompany it and make its path more gracious without compromising its effect. Fundamentally, to add a dash of irony, if movement was the keynote of the period and "to follow a direct path" the slogan of women's emancipation, it is clear that one could go faster by means of the circle of the wheel than with the straight lines of one's own legs. And even faster by means of the union of circles and lines represented by a train running on its track.

The circle, or the arc of a circle, the ellipse, the barrel shape—these are the lines that Cartier returned to repeatedly during those years. This is seen in the three brooches much discussed here, as well as elsewhere in this book. Thus, the brooch on page 118 above, with its clean geometric line, has its upper half in the form of a strict triangle prolonged at the base like a Y. However, the predominating straight lines and acute angles only serve as accompaniment to the fall of precious pearls and,

especially, to highlight the enormous pink pearl, extremely rare, of 111.28 grains, round and alluring, which completes the descent. Or, even better, look at the corsage brooch, also known as a stomacher, on page 123, shows that the play of linear and curved lines, almost inexhaustible in its continuous rebound from ring to ring, from segment to segment, is not limited to its contours but is repeated in the various shapes of the gems: rose-cut or brilliant-cut diamonds, rounded or square, pear-shaped or spherical emeralds. Sometimes the design is angular and rectilinear while the gems or pearls are round or drop-shaped. At other times, gems cut in a strict square, which might attain a considerable weight, are enfolded by a design with softer contours (as, for instance, the necklaces, necessarily curvilinear).

All of these jewels, of course, are mounted on platinum, infrequently combined with gold. Starting from the lightness and discreet luminosity of the precious metal, the designers of the Maison Cartier could give free rein to their fantasy, creating a scaffold of simple or complex geometric constructions in their unceasing quest for original forms.

With its chromatic neutrality, it proved especially suited to the first bold color combinations. However, color is a subject that deserves a chapter all its own, as it has a deeply rooted history and its use led to particularly felicitous results.

Design for a "tutti frutti" bracelet in platinum, emeralds, rubies, sapphires, onyx, and diamonds. Archives Cartier Paris, 1920s

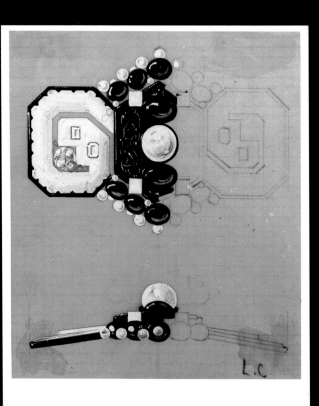

Design for a brooch in platinum, emeralds, diamonds,
pearls, and enamel. Archives Cartier Paris, 1925

THE SOURCES OF PLATINUM

Excavations in Zandsloot led to the discovery of platinum deposits in South Africa. The photograph appeared in Mining & Industrial Magazine on 22 December 1923.

*N*ot counting the marginal sources of platinum, such as the United States, Indonesia, and Australia, all the platinum mined and used since its discovery come from the reserves of four countries—Colombia, Russia, Canada, and South Africa—each of which has in turn held the role of its leading producer.

Until 1825, when the Russian deposits in the Urals began to be mined, platinum came exclusively from South America, then in the hands of Spain. The Colombian mines were not organized around commercial lines until the second decade of the twentieth century, by which time their deposits were already on the verge of giving out. Colombia had become a producer of secondary importance. The country that had supplied the entire world production for centuries today holds reserves estimated at about 0.2 or 0.3 percent of the world's deposits.

Starting in 1825, platinum discovered throughout the length of the Ural Mountains began to replace South America as a source. Like the Colombian metal, it came in the form of alluvial deposits. The Russian czar immediately nationalized the site, but production remained under the control of a few large English, German, and French companies that intended to satisfy the world's entire demand for platinum, which had soared. These deposits averaged 95 percent of world production every year until World War I.

The war and the Russian revolution precipitated a crisis in production and skyrocketing prices, until the new Soviet government undertook a systematic study to locate the areas where the metal transported by the waters originated. Their research paid off handsomely; platinum, along with nickel and copper, was discovered in the extreme northwest of Siberia, a thousand miles beyond the polar Arctic Circle in a region where the average temperature hovered around -30° Centigrade. Even under such severe climatic conditions, the mines of Norilsk still furnish today 90 percent of the Russian production of platinum and other metals of the same group as by-products of the extraction of copper and nickel.

Meanwhile, a new producer appeared on the market at the beginning of the twentieth century: Canada. In the province of Ontario, an enormous meteorite that fell to the Earth some two billion years ago had created the Sudbury basin, breaking through the Earth's crust to form outcroppings of magma rich in minerals, including the platinum metals. The industrial exploitation of the deposits began in 1901, and by 1934 Canada already provided 50 percent of the world production. It maintained its lead until the middle of the 1950s, when its mantle passed to South Africa.

In 1924 a platinum "fever" exploded. The German geologist Hans Merensky had located a platinum-bearing lode just beneath the ground on a farm in Transvaal. The vein proved to be the richest and most important deposit in the world. Following its course, the strong presence of platinum metals was discovered in the area of Rustenburg, which soon became the world capital of platinum. Dozens of mining companies appeared, as well as commercial ventures encouraged by speculators, who seemed to have transferred their traditional fascination with gold to the new metal. However, the extraction and refinement of platinum presented many more problems than those of gold, calling for larger and longer-term investments. When platinum fell to its lowest price ever, following the stock market crash of 1929, its collapse swept out all the mining ventures except for the two most stable, which then merged to create the firm Rustenburg Platinum Mines, the most important company in its field in the world.

From the 1950s, South Africa became the world's greatest producer of platinum. It currently produces nearly 77 percent of the world's production, about 104 tons a year. The mining companies carefully regulate this production in order to maintain a constant balance between supply and demand and, therefore, ensure the stability of its price. Platinum had come a long way from the days when it was used to adulterate gold: today, the price of the former is 20 to 30 percent greater than that of the latter.

Almost 80 percent of the entire world production of platinum comes from South Africa. The metal is concentrated in a vein less than a foot high called Merensky Reef.

THE COLORS OF THE EAST
1926–1939

Different concepts coexisted in Cartier's production of the 1920s and 1930s, with white Art Deco and abstraction on the one side and figuration on the other. This is readily seen in a comparison of three Cartier creations, all from the same period (the first two date from 1930, the third to 1929) and belonging to the same type of jewelry, as all three are bracelets.

On the bottom of page 154, is a perfect example of Art Deco in white, composed of platinum, diamonds, and rock crystal. Its line is strictly geometric, with a succession of eight identical decorative elements in an angular shape, connected by even more angular links. Its dazzling light arises from not only the reflective transparency of the rock crystal and the rarefied and discreet supporting platinum but also from a considerable number of diamonds, more baguette-cut than brilliant-cut. Whiteness, luminosity, geometry: this is the height of abstraction and coolness, the exemplification of the straight line. To introduce a note of softness and the sensual allure of a curve, one would have to imagine this bracelet around a woman's wrist. Photographed flat, as it is here, the jewel resembles a frieze.

On page 156, a very different bracelet is shown, with its warmth. The green of the emeralds, the blue of the sapphires, the red of the rubies sketch vivid forms around the cold brilliance of the diamonds, which they in turn absorb and reflect. The variety of the gemstones, of their shapes (leaf-shaped or spherical), and of their surfaces (smooth or engraved) compensates even for the inevitable flattening produced by the photograph and registers the full three-dimensionality and lushness of the image. It is no coincidence that such jewels received the amusing name "fruit salads," which was turned into "tutti frutti," the name they are still known by today throughout the world. Nature, held at arm's length by the sublime abstraction of Art Deco, made its way back if not in the shape of living beings, at least in the vitality of colors and the sensuality of volumes. From strict geometry and its extreme purity, which favored the intellectuality of formal research, design had moved—the same year, and without a visible transition—to the luxuriance of creation. It wrenched the senses from their lethargy—not only sight, whose craving for color was fully satisfied, but also touch and taste, aroused by a subtle temptation, a sudden urge to caress and savor.

The chimera bracelet on page 168, made a year earlier than the other two, actually offers a further development. A round, flexible bracelet, it is composed of a circular frame in platinum in which diamonds are closely set along the tight links of the frame. The lower half of the circle could be a classic example of white Art Deco in which the curve took over from the straight line. But this lower half circle, which progressively widens from link to link, finally reaching its full size at the level of the horizontal diameter, is suddenly interrupted by two rectangular bands of sapphires, and above them two chimeras that close the bracelet raise their heads and confront each other with gaping jaws. The chimera being a mythical creature, this is not a case of the imitation of nature, yet the two monsters are rendered in realistic detail. The gems are raised to form manes, tapered to form teeth, and ridged to create volumes that evoke living tissue. Two cabochon sapphires represent the bulging, forbidding eyes; calibrated sapphires form the nose and ears; cabochon

emeralds the jaws and wings—the shape and color of the gems seem to have been selected to create a realistic effect rather than a stylized one. Looked at in this light, the lower semicircle loses its semblance of purely abstract geometry and takes on the appearance of the tails of the reptiles linked together.

Considered together, these three bracelets suggest the range of production in this period. Now, Art Deco was at the height of abstraction and geometric purity in its reach for the spiritual light of pure white; and plants and animals teemed exerted their irrepressible sensuality and allure. By the end of the 1920s (as demonstrated by the Cartier jewels singled out above) the pull of Art Deco began to diminish, while an eclectic style (which, except for rare periods when it was more or less repressed, had been present throughout the entire century) came forth with renewed vigor. Art Deco was one of those rare artistic movements capable of restraining these various and far-ranging curiosities. It did so by enclosing them, almost symbolically, in a strict geometry, transmuting the contradictions of the century into a regular polygon, evincing an illusory unity.

A main influence in twentieth-century Europe was the impact of countries previously subjugated in the nineteenth century. At that time, the Europeans had heavily colonized Africa and the East, but in the twentieth century, Africa and especially the East colonized the European soul. India and China, in particular, where color dominated over form and feeling over rationality, introduced into the core of Europe's economic and military triumph the germ of psychological and cultural defeat.

In the travel narratives of the end of the nineteenth and the beginning of the twentieth cen-

The maharaja of Patiala with a turban ornament and necklace specially made for him by Cartier.

Design for a drop brooch in platinum, emeralds, rubies, sapphires, and diamonds. Archives Cartier Paris, 1928

turies—the period when this genre enjoyed a huge success in Europe—color prevailed over every other element. The East, by definition, denoted color in its markets and temples, animals and landscapes, flowers and fruits. As well, it conveyed adventure in its immense spaces and seeming regression in time. The East also signified eroticism, a perception that the West had formed with the "Thousand and One Nights" almost two centuries earlier. It possessed a wondrous balance that distanced it equally from reason and brutality—the two destructive poles of European eroticism—while it united, with supreme refinement, the animal grace of the body and the specifically human sensibility that is the expression of culture. These were the elements of fantasy and escape that the European public sought in travel books and novels, and, doubtless, found.

However, just browsing through the pages of a "light" writer like Pierre Loti, not to mention a great traveler such as Blaise Cendrars, so much more rebellious and openly subversive, shows to what extent the journey to the East is never free from a certain pain on returning. Even

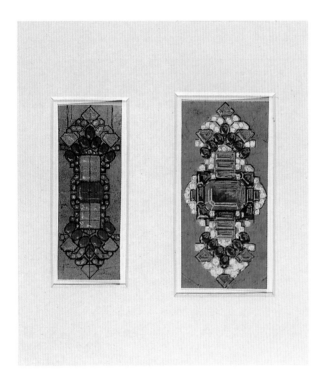

Designs for two brooches in platinum, diamonds, and light and dark citrine. Archives Cartier London, 1930s

when the journey takes place in literature it becomes both real and personal. The impassive East proved irresistible for Westerners.

Thus, color invaded France not only as a form of chromatic expression but also as a physical and psychic restlessness of movement and exploration. This duality surfaced with remarkable clarity in the experiments of the major artistic movements of the end of the nineteenth century and with special power in the beginning of the twentieth century, in the work of the Fauvists. Led by Henri Matisse and including Maurice Vlaminck,

and, in its early phase, Georges Braque, the Fauves—which means "beasts" in French, endowed by a French critic—made their name at the Salon d'Automne of 1905, where their violent use of pure color scandalized the critics and the public. The canvases of the Fauvists provide a strong example of the way in which contemporary art at the beginning of the century began to reexamine form, its basic constituent. Color, in the works of the Fauvists, overwhelms form. This was not static color, placed inside precise and well-defined contours, but turbulent color, fiercely dynamic, pushing at the boundaries of drawing, color thrown on the canvas in the form of spots and whirls, creating the effect of a vortex.

In addition, in the second decade of the twentieth century the East "came" to visit Europe and its uncontested cultural capital, Paris. It came on the tips of its feet and in the costumes of Russia. Starting in 1909, at Paris's Théâtre de Chatelet, the unforgettable reign of Sergey Diaghilev's Ballets Russes began, which remains a key event in the intellectual, artistic, and cultural history of the twentieth century.

Diaghilev, impresario of exceptional talent and a consummate artist with a lively, limitless curiosity, offered superb spectacles with the best dancers of the day (Vaslav Nijinsky, Michel Fokine, Ida Rubinstein, Anna Pavlova, Tamara Karsavina), the music of the most brilliant contemporary composers (from Claude Debussy to Richard Strauss, from Igor Stravinsky to Sergey Prokofiev, from Erik Satie to Francis Poulenc), and decors conceived and realized by such premier artists as Léon Bakst, André Derain, Alexandre Benois, Picasso, Giorgio De Chirico. European art, at a moment of its most splendid flowering and experimentation, lent its instruments and its paintbrushes to the blaze of Eastern colors. And the

The actress Gloria Swanson wears a pair of bracelets in platinum, rock crystal, and baguette-cut diamonds made by Cartier in 1930.

East seemed to repay European art by giving new life to its explorations and new wings to its dreams.

Always attentive to different trends, always ready to anticipate the next fashion, Cartier understood that the Eastern craze, both satisfied and reawakened by the memorable entrance of Diaghilev and his Ballets Russes, had paved the way for a progressive revolution in terms of materials and color in the traditionally conservative world of jewelry.

Since the beginning of Art Deco, a freer and more inventive production had coexisted with the restrained and severe black and white jewelry (platinum, diamonds, and rock crystal with black enamel and onyx) in the Maison Cartier. From within the strict geometry of its design, unusual colors and materials began to vibrate. Emeralds, sapphires, and rubies, of course, but also turquoise, coral, jade, lapis lazuli, pearls and mother-of-pearl, as well as colored and opalescent enamels, introduced greens, blues, reds, and also the lustrous pinkish white of pearls. Also employed were the fleshy, orange tones of coral, itself a disturbing material that exists in a sort of limbo: animal by nature, vegetable in form, mineral in the way it is worked and the way the final product looks.

The avid eclecticism of the contemporary period, of which Cartier, with his unlimited curiosity, was eminently aware, clearly exerted its power even on the rational and abstract heart of Art Deco from the first. More to the point, the use of different materials naturally culminated in unusual chromatic combinations, starting with the very new, even scandalous, union of green and blue, which would become emblematic of Cartier's daring and originality—sapphires and emeralds for jewels, lapis lazuli and nephrite for accessories. The association of green and blue seemed to evoke the theme of the East as well as of the voyage, which held Europe in thrall during those years: the blue of the crossing, the green of its goal; the blue of the ocean, the green of a luxuriant nature; the

blue of space and thus the faraway, the green of the Earth and thus the mooring. These are the symbolic colors, in the form of sapphires and emeralds,

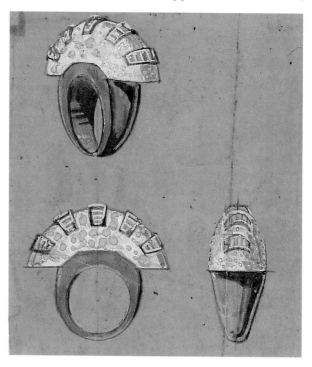

Design for a ring in platinum, pavé diamonds, and baguette-cut diamonds.
Archives Cartier Paris, early 1930s

that spring from the European contradiction: the blue of rational abstraction and the green of life's instinctive force.

From the first appeals to the senses evident in certain of the creations made between 1915 and 1925, still very marked by Art Deco and its geometry, to the savory success of the "tutti frutti" jewelry, the East would also play a different role from the traditional one of simple polar opposite: the direct and unprecedented role of patron. In this adventure, platinum played a major role as well.

Its whiteness made it the ideal setting for diamonds in Garland style jewelry, but its success as the setting for colored gems was less predictable. The experiments of the first period of Art Deco demonstrated the metal's extraordinary ductility, not only physical but also aesthetic. Its brilliance, cold in association with diamonds, grew warm when combined with color. The fact that it can be made into very fine wires rendered it practically invisible and, at the same time, permitted the elaboration of complex geometries on a light base, in terms of weight as well as volume.

The beauty of Cartier's creations that combined platinum with a variety of materials struck the East like a counterblow, much as the colors of the East had seduced Cartier. The great Indian princes, who possessed treasure troves of precious gems and wore fabulous parures of inestimable value, were strongly attracted by the novelty of settings in platinum, which seemed to enhance colors of gems better than gold. This was true despite the symbolic value of gold, the metal of the sun, in these faraway lands. Adopting platinum, however, carried with it the pleasurable sensation of being at the peak of European taste. The maharajas, naturally, were at home in England, being at once the subjects and the instruments of British power in the Indian colony. The children studied in the best English schools, and London had become their second home as well as their capital. The Cartier shop on Bond Street, one of their favorite places to shop, from the end of the 1920s received orders from numerous Indian princes who desired new settings in platinum for their most amazing parures. In 1929, the jewels of the maharaja of Patiala, given a new setting and transformed by Cartier, were exhibited in Paris, in the shop on the rue de la Paix, before being restored to the fabulously wealthy client. In this extraordinary exhibition, which stole the hearts of the Parisians, converged all the opposing forces that arrived from the East and in returning there jolted the geometric vein of Art Deco, restrained and pure, into the unexpected abundance of color and life in the "tutti frutti" creations.

From India to China, and from imaginary fruit to mythological animals, the path was short

but far from linear. Logically, China, with India and the Islamic world, was one of the poles of attraction for European restlessness, one of the dreamlike and evanescent symbols of the unknown East, perhaps even more magically and mysteriously. But the connection was not as direct as it might at first seem.

In reality, the works that incorporated chimeras and dragons, pagodas and Buddhas, the

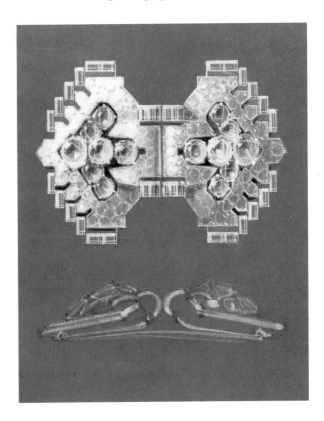

Design for a brooch in platinum and diamonds.
Archives Cartier Paris, 1936

mythic and sacred emblems of China, demonstrate even more clearly than the fruit salad pieces an unarguable fact: the formal rigor of Art Deco from that point on was only a cage in which the jeweler of creative imagination and universal curiosity had no intention of remaining prisoner. Cartier himself, with his precocious intuition, was among the first to set the terms of Art Deco's domination and development. Yet Louis Cartier was also the refined incarnation of the insatiable and multiform eclecticism that was the twentieth century's primary, if undefinable, characteristic. Abstract and geometric rationality was too narrow for him; it imprisoned his creative genius and his bent for diversity.

Beside the creations in the realm of jewelry and watchmaking, which by then was part of the company's traditional production, everyday objects reinterpreted through the art of the jeweler were increasingly offered to the clientele of those years. Accessories for ladies (some admirable examples can be found on pages 150, 166, and 167), small handbags for evening (page 157), cigarette cases (page 164), not to mention the famous "mystery clocks" (pages 165 and 173), from this moment on formed part of the Cartier oeuvre. All these objects adapted to the Art Deco style magnificently, which, with the clean geometric lines of its design, proved ideal to confer elegance on all forms. But in order to transform them into veritable jewels, Cartier drew freely on all possible repertories for the decorative elements, including a real and also a mythological bestiary. The encounter with China proved especially fertile in this respect, and it gave birth to several series that remain famous today, as, for example, the chimera bracelets or the pagoda desk clocks. But Cartier also drew on the repertory of real creatures: parakeets, insects, fish, and, especially, cats.

Beginning in the 1930s, the explosion of Art Deco was more dramatically emphasized by the rift between contrary poles of rationality and irrationality, between the abstract and the concrete, between science and imagination.

As the 1930s began, the euphoria of the Roaring Twenties had already been dealt the fatal blow of Wall Street's Black Thursday (24 October 1929); in the current reality, the Great Depression constituted the somber end of a dream. Soon the rise of Nazism (1933) would cast even blacker shadows over the future of the world.

The economic crisis of the 1930s marked a relative decline in platinum, now having become too expensive and refined. Cartier continued to use it for the top of his line, as illustrated on pages 152–53 and 174–81. These were fabulous creations, many of them featuring stones of exceptional size. These inventions were bound for those clients who had no need to fear the period and its threats of penury, including, increasingly, the great movie stars.

Movie stars, in the climate of disillusion of the 1930s, represented the last surviving dream—the last, but of tremendous vitality. The invention of the talkies in 1928 brought about a veritable revolution. In the years that followed, American movies, aided by an influx of great German and Austrian professionals fleeing Nazism, developed strong screenplays and direction that long remained unequaled. It was also the cinema, more than the novel in those years, that effectively depicted the social contrasts of the period, and it lent its voice to the celebrators of life as well as to the disinherited.

Dreaming is a physiological necessity that every human being must satisfy in one way or another. Thus the great leading ladies of Hollywood, who instilled dreams in untold audiences, needed to find a dream in their turn. Significantly, they dreamed of a parure made by Cartier, especially one in platinum and diamonds. Platinum's success in Hollywood was summed up by the title character in Frank Capra's *Platinum Blonde* (1931), played by Jean Harlow. This was the origin of the myth of the platinum blonde, which, after Harlow herself, had various incarnations, from Marlene Dietrich to Marilyn Monroe, across a quarter of a century and a world war.

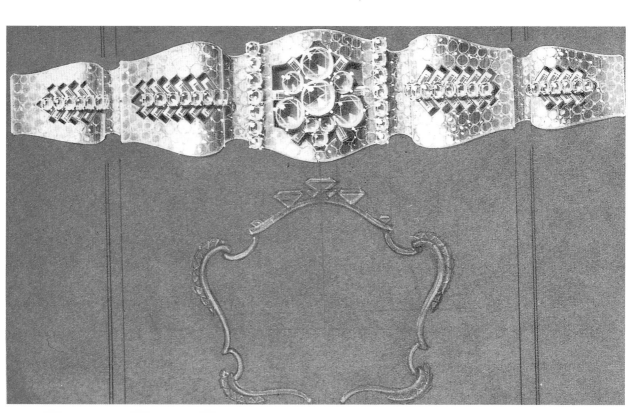

Design for a bracelet in platinum and diamonds.
Archives Cartier Paris, 1938

N° 1

£670. Special Cost

Design for a necklace in platinum,
diamonds, rubies, and emeralds.
Archives Cartier London, late 1920s

THE PROPERTIES OF PLATINUM

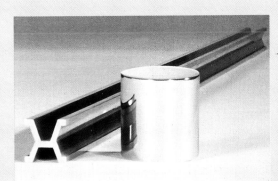

Platinum is impervious to most outside agents as well as time. For this reason, Marc Étienne Janety selected platinum for the standard kilogram and meter.

Platinum as a material cannot be defined without using superlatives: it is the rarest, the purest, and the most precious of all the metals. It is lasting, resistant to chemical reaction with most reagents, including oxygen and acids, except aqua regia. It will not melt at high temperatures, yet it is ductile. It is extremely hard, yet it is highly malleable. Above all else, platinum is rare. Today, only 4.3 million ounces (about 135 tons) are extracted every year from around the world, against over 80 million ounces of gold (about 2,700 tons). If all the platinum extracted from the earth since the beginning of its history to the present were collected in a single lump, the result would be a cube of about 15 feet in each dimension—about the volume of an average room.

Platinum, especially in the form used for jewelry, is extremely pure. Its properties allow it to be worked at very high levels of purity, which is not true of the other noble metals. Platinum jewelry generally employs alloys with a proportion of 95 percent, as opposed to 75 percent in the case of 18-karat gold.

Moreover, platinum is heavy. Its specific gravity (21.45 grams per cubic centimeter) is among the highest known, far higher than that of silver (10.5) or gold (19.3). Even taking the difference in purity into account, an object made of solid platinum weighs more than one of solid gold in the same dimensions, a difference that can be felt by holding the two objects in the palm of each hand.

Extraordinarily heat-resistant, platinum must be heated to 1,772°C. before it will melt, against 1,063° for gold and 961° for silver. Only most of the other platinum metals have a higher melting point, notably osmium, which melts at 2,700° C. It is, in fact, this high melting point that has made platinum difficult to use, even several centuries after its discovery, as it cannot be forged or worked. On the other hand, its very low heat conductivity (73 watts per linear meter for each degree Centigrade, compared with 293 for gold and 419 for silver) makes it virtually impervious to heat. For this reason, an alloy composed of 90 percent platinum and 10 percent iridium, a metal from the same family, was chosen for the official standards of weights and measures of the metric system at the end of the last century.

Platinum resists not only heat but also chemical reagents. It will not react at all with oxygen and few acids can make a dent in it, which led to its use, since the middle of the last century, in making laboratory utensils, such as crucibles, in which sulfuric acid is combined with other compounds. For different reasons, this property is also appreciated by jewelers, as its inalterable nature represents a fundamental guarantee for the settings of precious stones.

Platinum's ductility has also been put to use. Once it is melted, without losing its strength or resistance to corrosion, it can be made into sheets or extremely fine wire. In this, it is alone among the metals.

The unique luminosity of platinum makes it ideal to showcase diamonds in jewelry settings. Unlike all the other metals, it does not detract from the natural fire of the diamond, whose brilliant light constitutes the singular beauty of the stone. By the same token, platinum provides a natural white that the artifically produced white gold cannot match. An alloy of gold and silver, white gold was invented by a German goldsmith to satisfy an increasing demand for a white metal in jewelry, a demand that platinum could not then meet as it had been declared a strategic material by all the warring countries at the beginning of World War I.

Platinum is the head of a family of dense metals known as the platinum metals—iridium, osmium, palladium, rhodium, and ruthenium—no less rare and precious than their noble "parent." They, too, possess special physical and chemical properties that make them highly desirable for various industrial and scientific uses.

Even with its wonderful luminosity, platinum is discreet, with an elegant, restrained color favored by men for their luxury accessories, especially watches.

Platinum is extremely ductile. A gram of the metal will yield a very strong thread over a mile long. Such thread could be used to make fabric, as for this remarkable wedding dress made in Japan.

FLORA AND FAUNA
1940–1967

Born under the sign of a faith in scientific and technological progress and in the supremacy of reason, the twentieth century did not even reach its midpoint before it saw its myths destroyed and its bold constructions demolished. World War II swept away the remainder of the utopia that World War I had left intact. The smoking ruins of European cities destroyed by bombing were also the ruins of reason.

The immediate postwar years have entered history as those of the reconstruction—an ambiguous and, in a certain sense, misleading word. Europe, with help from victorious America, rapidly cleared away the rubble, and in a few years cities came back to life, factories restarted production, national economies found new breath, political systems reorganized, and well-being returned almost everywhere. In general, when writing the history of Europe from 1945 to the middle of the 1950s, the use of the prefix *re-* could suffice. One could then also speak of "re-construction," but it does not take the wisdom of Confucius to realize that nothing can be rebuilt exactly as it was after it has been destroyed—not a house, nor, even more logically, a city. Not to mention an entire continent. Europe was better at reorganization than at reconstruction.

The axis of history resolutely shifted westward. This fact was largely evident after World War I, but the phenomenon had slowed or been weakened by Europe's uncontested cultural domination. If the cultural axis, unlike the political axis, tended to turn at that time, it was toward the east rather than the west, and this had more to do with desire than reality. In fact, throughout the period between the world wars, American artists and intellectuals flocked to Paris, undertaking a journey of initiation whose value was largely symbolic.

Europe's cultural prestige did not vanish, of course, but it resided more in the past than in the present. The United States had become a leader with its great cities and products, and its life-style.

The Marshall Plan (named for George Marshall, President Truman's Secretary of State) was an American plan to aid in Europe's reconstruction. The price paid was the massive invasion of American products on the European continent and United States' hegemony in NATO, a military alliance of that country and Europe's principal nations.

When the world went to war, it was neatly divided in two—civilization against barbarism—and it also divided in two at the end of the war—capitalism versus Communism. When the first opposition made way for the second, cultural ideals and goals disappeared for the exclusive benefit of economic interests and ideologies. The world of art and intelligence, which had found itself united against the Fascists and Nazis, divided evenly between the two new camps. It was no longer even possible to say where culture itself actually resided; what was clear was the division of the world into two blocs, whose line of demarcation cut Europe in half with a sublime indifference with respect to people and cultures, just as that continent had divided the world over the centuries by drawing straight lines on its maps. On one side or another of those lines, natives of South America in the seventeenth century or of Africa in the nineteenth might suddenly find themselves Portuguese or Spanish, English or French. On one side or another of a more tortuous line, which quickly

took the name Iron Curtain, and was even more impossible to cross. Europe in the second half of the twentieth century found itself the citizen and free trader of a popular democracy.

The postwar European cultural climate is easily summed up by two important movements that dominated the scene in the 1950s, one concerning philosophy, literature, and ethics, the other more specifically artistic: existentialism and Art Informel. Both of them (especially the first) bear witness to two contradictory tendencies: on the one hand, their impact demonstrated that Europe, even in the grip of a profound crisis of values and identity, could still produce important cultural movements and elaborate intellectual thought; on the other, because of their basic pessimism, they proved that European culture had definitively abandoned the optimism of the beginning of the century and the euphoria of the immediate postwar period.

Existentialism, with Jean-Paul Sartre as its leader and embodiment, is a philosophical doctrine that points out the absurdity of living. It maintains that we are thrown into the world like groundless strangers but are constrained to maintain continuous and extremely close relations with it. Placing itself in direct polemical opposition to traditional philosophical thought (especially that of an idealistic nature), which would found a "science of being," existentialism denied all possibility of defining the essence of humanity and concentrated its attention on existence, pure and simple. The experience of daily life—the sensation of emptiness that accompanies the effort to play one's part in relation to a world devoid of sense—is at the heart of existential thought, whose key word is anguish.

*The Duchess of Windsor wears
a tiara, convertible to a necklace, in platinum,
diamonds, and thirty-nine emerald beads,
made by Cartier in 1940.*

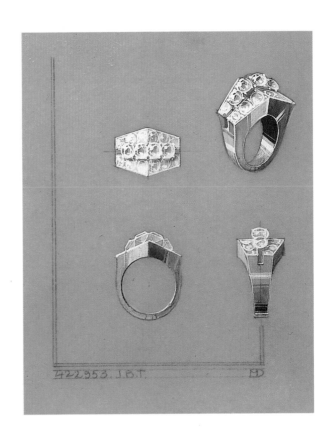

Design for a ring in platinum, white gold, and diamonds. Archives Cartier Paris, 1941

This dramatic vision of humanity confronted with its abandonment, without god or consolations, reverberated powerfully in many of those who had just emerged from the horrors of war. (The sole positive aspect of existentialism—its only alternative to the anguish of emptiness—was Sartre's proposal of "political action" or "engagement" in the service of an ideal cause.) Although the basis of existentialism had been laid down in the 1930s, it was not until after the war, in the 1950s, that the movement spread beyond a narrow intellectual domain to become a keynote, even a commonplace, of the period.

In its literary incarnation, existentialism found its most original voice not in Sartre but in Albert Camus. His novels, such as *L'Étranger* (The Stranger), *La Peste* (The Plague), and *La Chute* (The Fall), reveal better than any other work of the period the yawning chasm that separated humanity in 1950 from that of 1900. At the heart of his work is the theme of absurdity, the irreducible and paradoxical opposition between the irresistible straining toward an ultimate meaning of life and the flat assertion that no such thing exists. To the latter, the logical response would be the refusal of life. To the contrary, Camus's characters discover in the celebration of their "marriage" with the world the reason for their passionate attachment to life, even a love of life. It is significant that this marriage has nothing to do with the union of the individual with human society or history, nor (as Sartre proposes) with a political or social ideal. The only power capable of breaking the vicious circle of the absurd (the search for a meaning that cannot be found) resides in the secret beauty and harmony of nature. Camus, a Frenchman born in Algeria, spent his childhood on the banks of the Mediterranean, a source of so many of his ideas. Perhaps no other writer of this century went further than Camus to meld narrative and thought or employed a more limpid language and more classic style.

The fundamental themes of Art Informel (which paralleled the New York school of action painting, whose foremost and most famous proponent was Jackson Pollock) involve a refusal of reason, a marked mistrust of knowledge. This movement, which emerged in the 1950s, was based on the total demolition of form as a component of the pictorial composition. Even the concept of a figurative versus a nonfigurative painting (or classic versus abstract) was rejected. Nonfigurative, or abstract, painting, while rejecting the traditional vision of art as representation or reinterpretation of reality, apparently still made form the very center of its explorations and its mode of expression. One has only to think of Piet Mondrian, Wassily Kandinsky, or Paul Klee to confirm that the abstract painters placed formal investigation above all other concerns. In contrast, the adherents of Art Informel made color the only component of pictorial space. However, speaking only of color is

misleading. In reality, color in itself did not interest these artists; what concerned them was color as material, as matter to manipulate. Paradoxically opposing itself to abstract art without resorting to figurative art, it thus confined artistic expression entirely to the most concrete component of the "action" of painting: its material substance.

It is illuminating to review the background

Design for a kingfisher brooch in platinum, gold, emeralds, sapphires, rubies, and diamonds. Archives Cartier Paris, 1941

of the period's pessimism as a prelude to discussing Cartier's production in the 1950s.

To the vicissitudes in the world, which certainly did not favor the blossoming of a highly stylized artistic season, could be added those of the family and the business. In the same year, 1942, two of the three Cartier brothers died: Louis, who had for close to forty years set the inspired level of

Barbara Hutton wears the emeralds of Grand Duchess Vladimir and the Pasha diamond.

creativity, and Jacques, the youngest, who was in charge of the London branch. Only Pierre, the head of Cartier New York, remained; he died in 1965. In Paris, Louis's close collaborator, Jeanne Toussaint, took over the artistic direction of the company. Louis's death marked the end of a period, but it also marked the beginning of another, both linked by the inexhaustible creativity of Cartier, which always anticipated the times and the fashions, always placing the Maison Cartier in the fore.

The Europe that emerged from the war had changed dramatically. For one thing, Western capitalism had not yet overcome the economic crisis set off by the stock market crash of 1929. Only the considerable investments in public works (President Franklin D. Roosevelt's New Deal is the most obvious example) and the subsequent re-orientation of production to a war economy had salvaged the national economies and slowed unemployment. But among all the Western countries, these remedies presupposed the increased intervention of the state in the economy, both in the public and private sectors. This became intensified in the postwar period when all the countries of Western Europe (led by France and Italy, but also Great Britain and West Germany, to cite only the most important and most heavily populated) saw a period of heavy nationalization of public services and even some of their basic industries. The middle classes of these various countries were ill-served by these changes until the beginning of he 1960s, with the decision to found the European Economic Community, which led to the internationalization of capital, of business initiatives, and of markets.

The ruins of war, the expansion of poverty, the needs of reconstruction called for sacrifices from everyone, or almost everyone, and restraint in spending even by those who had means. Later, with the return to economic growth, came a phase of the redistribution of wealth, spreading a better standard of living to a greater proportion of the population but leaving the wealthy individual with less money. Overall, the number of wealthy individuals increased, but their individual worth shrank.

On the whole, it was not a propitious time for jewelers, as much on the level of creativity as that of business. Cartier held its own by drawing on its past and the legacy of ideas that had yet to be exploited. For Cartier, the prefix *re-* signified abundance.

In the preceding chapter, it was pointed out how Cartier, at the height of Art Deco, began to alternate its strict geometric lines with the colors and forms of nature. That vein became prodigiously productive from the 1940s to the 1960s, constituting the most original aspect of Cartier's production in a period when originality proved to be a rare commodity.

Flowers, above all, and leaves became more significant sources of inspiration (seen on pages 184–89, 214–15, 217, 220, and 221). Stylized nature has yielded to figuration. The roses of this period are true roses, the palm trees real palm trees, the leaves actual leaves. Without resorting to simplistic imitation, the jeweler adopted a resolutely naturalistic approach, with the intention of restoring all the harmonious asymmetries of a living flora by means of the elegance of the design, the precise definition of volumes, and the judicious choice of colors. This was very different from the purified stylization of the Garland style: the contours, while clear-cut, are far from simple. It was equally removed from the scrupulous respect for symmetry. For example, look at the bouquet of flowers at lower right on page 185, or the palm trees on pages 214–15: it is clearly impossible to draw a vertical axis that will divide these objects into two halves of equal visual weight, in terms of design as well as volumes. On the other hand, the new style also had little to do with the "tutti frutti" style, which had no other goal than to evoke sensuality from the abundance of its plant forms. Here, the volumes do not suggest forms, they faithfully reproduce them; the colors do not stimulate the appetite, they satisfy the sense of sight; the desire to touch and taste has given way to a vague desire, barely marked, to take in. Jewelers no longer made the flora bend to formal requirements, they now paid homage to it with the instruments and materials of their art.

The naturalistic approach was perhaps even more telling when it went from flora to fauna. To the harmony of forms was added the illusion of movement. The birds on pages 194–95 are clearly represented in flight: their twisting necks, the almost greedy bent of their beaks, the airborne lightness of their wings are artifices devised by the jeweler to set these birds free from their simple ornamental fate as brooches fastened on a woman's dress. And what can one say of the famous panther on page 210? The different positions of its paws, lightly separated in front, tensely crouching in the rear, the enveloping curve of its tail, the very expression of its face, somewhat wary and concentrated, make it a large cat ready to play and transform the enormous cabochon sapphire of more than 150 carats into a ball larger than the cat, which attempts to hold it between its paws; clearly, it cannot succeed for long, as the ball cannot hold still.

This naturalistic phase at Cartier was in complete accordance with the personal taste of Jeanne Toussaint, who inspired the stylistic choices of the jeweler at the time, but it was equally in tune with the needs of the period. Like Camus's characters, the postwar individual found in nature not only an incomparable source of comfort but also one of the few surviving sources of good taste.

As this was a period of reconstruction, it should have corresponded to a golden age of architecture. Entire cities needed to be rebuilt on their ruins, while an unprecedented population explosion made the conception of entire new neigh-

Design for a flower brooch in platinum, gold, diamonds, and sapphires. Archives Cartier Paris, 1943

borhoods imperative. However, architecture seemed to have completely lost the idealistic verve that had animated its experiments in the first half of the century. Rationalism, functionalism, modernism, these were some of the schools of architecture that, in the early decades of the twentieth century, tried to celebrate the totally new union of industrial development and the increasing quality of life and the habitat. Great architects like Le Corbusier and Henry van de Velde, Walter Gropius and Frank Lloyd Wright took advantage of technical advances, new materials, and the possibilities offered by mass production to propose a new harmony between the individual and the environment.

After the war, this liberal approach seemed to have run out of steam. Those still pursuing it were the same ones who had initiated it, and they had become old (the four aforenamed masters all died between the end of the 1950s and the beginning of the 1960s). Le Corbusier still had such masterpieces to offer as the chapel of Notre Dame du Haut at Ronchamp, where he turned concrete into an expressive material, appreciated for its plastic properties and the ease with which it could be used for straight lines, round volumes, clipped right angles, as well as for its strength and its acceptance of color. But this was an isolated instance. While in the first half of the century architecture collaborated with industry to discover functional character and beauty in technology, after World War II the building became the instrument of industry, which propagated in a servile manner the least costly innovations.

European cities got new faces or, at least, when they were lucky enough that their historic centers escaped the war's ravages, new sheaths. The new residential areas enclosed the vestiges of art and history in their ugly embrace. The inability to build according to a human aesthetic perhaps sums up Europe's postwar disarray better than any other degeneration of style.

Other impulses, which would become increasingly emphatic over the years, progressively led the European to reevaluate nature. Ecology and the flight from the cities had not yet touched the 1950s and 1960s. However, for the first time in the history of humanity, a totally new fear materialized—the threat of nuclear conflict. For the first time, the idea that the world could self-destruct, that Homo sapiens could play a fatal role in the planet's destiny, began to take hold. It prepared the heart of Western man for that painful self-criticism of the development of his own civilization, and it would remain a part of his daily life. The return to nature was a major component of this new feeling; paradoxically it would unleash massive destruction on the environment due to an excess of tourism and the crazy race to secure second and third residences.

It is enough to glance at the jewels mentioned above and on the captions that accompany them to affirm that after World War II the use of platinum became more prudent than it was during the preceding decades. In most of the nature-inspired objects cited, it appears in association with gold, unlike its use in the creations of the Garland style or Art Deco. Platinum was still reserved as the exclusive setting for gems, especially diamonds, while gold regained its traditional function of defining lines.

This change is undoubtedly related to the fact that postwar jewelry presented a more circumspect luxury than that of the preceding periods. As was already pointed out, fine jewelry, whether due to reduced finances or to opportunism (ostentatious wealth was unseemly in this atmosphere of austerity), was forced to observe a certain economy of materials. The size of the gems was reduced and the surface of the "naked" metal increased. Moreover, gold was easier to work for extended surfaces than platinum, which was restricted to settings for stones. During the war, because of its multiple industrial uses, platinum had practically disappeared from circulation; some countries at war had gone so far as to forbid its use for nonstrategic purposes. So the jewelers had made their forms and designs "poorer," using fewer gems, a practice continued even after the war ended, due as much to inertia as to necessity.

A few very fine examples of jewelry that Cartier continued to make furnished an exception to this rule, but these were produced on commission. These orders clearly came from special clients, among them movie stars and the most glittering examples of high society. Among the first, Maria Felix, Grace Kelly, and Elizabeth Taylor were Cartier clients. Among the second, the American heiress Barbara Hutton (see the tiger parure on page 212) and, especially, the Duchess of Windsor (see the butterfly brooch on page 206 and the panther brooch on page 210).

To paraphrase George Orwell, one could say that the Duchess of Windsor constituted, within the bosom of an extremely narrow elite made up of unique cases, a case more unique than the rest.

Even Marcel Proust, the greatest bard of that curious phenomenon known as snobbery, could not have foreseen that the extraordinary transformation of Madame Verdurin into the Duchess of Guermantes (which concludes *À la Recherche du temps perdu* as an ironic prefiguration of the time to come) would be transcended in reality by the transformation of an American woman, no longer young, already several times divorced, into the "almost queen" of England. But this is exactly what happened in 1936, less than fifteen years after Proust's death: King Edward VIII abdicated in December of that year, several months after his accession to the throne, when it became clear that he could not gain approval from the Conservative government led by Stanley Baldwin for his marriage with Mrs. Wallis Simpson. In 1937, the ex-king, now simply the Duke of Windsor, finally led the woman he loved to the altar and made her, if not a queen, at least a duchess.

All in all, it can be argued that Edward made an intelligent choice. His love affair moved England and the rest of the world (among other reasons, because both had already passed the age

of forty, and few people at that age had sacrificed power for love), his marriage was a splendid success, and the means at their disposal for the rest of their lives allowed them to lead a much more carefree and sumptuous existence than they could have had as rulers. In addition, their passion for jewels assured them a posthumous fame no less inferior to that they would have acquired as king and queen in a period when the role of sovereigns had already become largely ceremonial.

The Duchess of Windsor, whose talent and drive spurred her to be more queenly than a queen, was one of the most elegant women of her day, and when it came to jewelry she literally dictated the fashion. It is through her that platinum, pushed to a minor role by the conditions of the time, regained its rightful and irreplaceable role in fine jewelry.

"After five o'clock in the afternoon, one should wear platinum, not gold." This law, in sum, was attributed to this refined duchess. Platinum as the ideal metal for evening jewels was an interesting novelty. Since gold is yellow, it pertains to the sun and the day; silver-white platinum belongs to the moon and the night. This distinction resurrects a metaphor as old as humanity: gold/sun and silver/moon. But silver, much less precious than gold, here is replaced by platinum, which is more precious than both. To contemporary humans, the night is, in effect, more precious than day. It is the free part of their time, the time for pleasure and adventure.

The Duchess of Windsor, like other of Cartier's important clients in the 1950s and 1960s (Barbara Hutton and Princess Nina Dyer, wife of Prince Saddrudin, son of the Aga Khan), was an ardent "sponsor" of the "cats" that Jeanne Toussaint loved so much, particularly the panther. But for those jewels created specifically for special clients, often conceived around gems of exceptional value, Cartier did not limit itself to the naturalist theme.

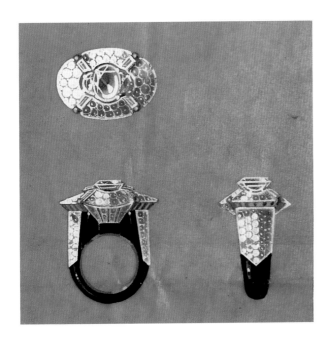

Design for a ring in platinum, gold, diamonds, and onyx. Archives Cartier Paris, 1949

As some of the pieces reproduced here demonstrate (those on pages 191, 202, and 203 are significant examples), Cartier began to draw freely on its past, yielding to a happy "autobiographical" eclecticism; thus, in the middle of the 1950s and 1960s, Cartier came out with bracelets and brooches in the "tutti frutti" style or Art Deco necklaces. In a period when style was in hiding, it made sense to revisit the styles of the past, particularly the recent past. In this way, Cartier anticipated the modern style that would soon spread throughout Europe.

One can already see, in this shrewd method of making use of its enormous legacy of experiments, projects, and designs, the foreshadowing of Cartier's ambitious renaissance to come several years later. The revolution of Les Must was knocking on the door.

Designs for two bird brooches in platinum, gold, white gold, emeralds, rubies, and diamonds.
Archives Cartier Paris, 1960–61

THE USES OF PLATINUM

The industrial uses of platinum are legion. From chemistry to electronics, from the automobile industry to petrochemistry and atomic energy, from medicine to aerospace, platinum plays a fundamental role in the improvements wrought by technology.

*S*ince the second half of the eighteenth century, when small quantities of platinum began to arrive in Europe more or less secretly, the material has been put to a remarkable variety of uses.

Its very high melting point, which set it apart from every other known substance, stood out as the most notable property of the new metal. As no one yet realized its noble and rare qualities, one of its earliest practical uses seems truly extravagant today: until the beginning of the nineteenth century, it was freely employed for the vents of firearms, that is, the opening in the breech of artillery in which gunpowder was ignited, because it resisted the intense heat produced by the explosion. The attitude that permitted such a use was not far removed from that of the Spanish conquistadors, who considered platinum a nuisance that contaminated the gold mines; they themselves substituted it for lead to make shot.

As soon as scientists began to appreciate the extraordinary qualities of platinum and the other metals of its group (all discovered between 1800 and 1810), it was reserved for more important uses. On the whole, from then until now, platinum has played a significant role in the areas of scientific research and industrial development.

To begin with, today's clean air depends on platinum; its catalytic properties make it an essential component of the catalytic converters in which the gases emitted from automobiles are transformed into non-noxious substances. This could be considered platinum's most important current application.

Platinum has also been instrumental in numerous inventions dealing with light. As far back in its history as 1813, Johann Wolfgang Döbereiner used it in his lighter, innumerable versions of which lit all the lamps and all the cigars in Europe before the match appeared. The filaments for the first incandescent light bulbs made by Thomas Alva Edison or Joseph Swan were of platinum.

In the production of energy, platinum is still considered a precious element, used in nuclear reactors and the refinement of hydrocarbons as well as for the manufacture of the electrodes of fuel cells. The earliest internal-combustion engines of automobiles employed platinum in the cylinder where the mixture of air and gas ignited. At the beginning of the century, F. W. Ostwald, a German chemist, exploited platinum's exceptional catalytic properties to create nitric acid in the laboratory. A fundamental invention in terms of industrial development, nitric acid also became a formidable instrument of death when it was used to make explosive devices after World War I erupted. But it is also among the basic elements of fertilizer, so integral to agriculture; even today, platinum remains essential to its production.

Because of its imperviousness to heat, platinum is used to make resistance thermometers, which measure extremes of temperature, both high and low.

One of the most compelling of platinum's uses, in terms of the future of humanity, must be in the medical field. A compound based on platinum has been in use for over twenty years throughout the world for its powerful effects against cancerous tumors. The amazing pacemaker that regulates the heartbeat also contains platinum. In this particular case, as in many other prosthetic devices, platinum is the material of choice for a reason both simple and mysterious: platinum is totally compatible with human tissue, so it does not cause allergic reactions or rejection by the body. This quality makes it irreplaceable in medicine and invaluable in jewelry.

Fiber optics

THE NEW ERA
From 1968 Toward the Year 2000

On 21 July 1969, hundreds of millions of people the world over, following the event live on their television sets, watched fascinated as American astronaut Neil Armstrong placed his foot, enclosed in special, very heavy boots of enormous symbolic weight, on the moon. This apparently small gesture gave rise to feelings, ideals, and myths so numerous that it is impossible to detail them without ending up with a long list.

The "conquest of the moon," from a cultural point of view, constitutes a magnificent paradox of contemporary man. Jules Verne's *From the Earth to the Moon* appeared in 1865, in a period when scientific positivism was at its height (note that Verne's novel falls exactly midway between Darwin's two fundamental works, *On the Origin of Species by Means of Natural Selection*, of 1859, and *The Descent of Man and Selection in Relation to Sex*, of 1871). Meanwhile, this imaginary voyage became reality a century later, when faith in scientific progress as an instrument of redemption and happiness for the human species was no more than a vague memory.

The "journey from the Earth to the Moon" was only the latest and most spectacular episode of the "race in space" that was one element of the Cold War. This latter, for which the moon, with its icy night whiteness, makes a perfect thermic symbol, is the only explanation for the enormous economic, scientific, technological, and human resources that the United States and the Soviet Union for about a decade expended on this "political" objective: to be the first to walk on the moon. The stakes in this game had nothing to do with intellectual or scientific curiosity and research and everything to do

with propaganda and the struggle for power. The best proof lies in the quarter of a century of total silence about what was then presented as humanity's magnificent astronautic destiny.

Rather than a prodigious future in which scientific progress would guide the human species toward far distant worlds, the conquest of the moon speaks of science's development as a tool of political or financial concerns.

As always in such cases, viewed from a distance, the meaning of this "race in space" was not a simple "sports" competition. In the deep and genuine emotion that we experienced watching the feats of Yury Gagarin and Armstrong could be discerned the symptom of a new ailment: the need for space, greater than the Earth itself.

One fascinating by-product of the conquest of the moon was the image it returned of the Earth taken from "outside" in photographs: unlikely phases of the Earth, crescents of the Earth, half or full Earths. It is disquieting to note the extent to which the goal of the voyage could be considered looking into ourselves, to give ourselves some distance from the world in order to see it better.

The 1950s and 1960s (two decades of economic "miracles" in many parts of the world, especially in Europe) addressed the inexhaustible richness of space, and the inadequacy between space and human desire. In 1969, the same year of the moon's conquest, Jack Kerouac died, barely forty-seven years old, the most emblematic writer of what is called the "beat" generation and the author of a novel that vividly sums up its themes and myths. *On the Road*, a title that could not be

more meaningful, is the bible of an entire generation that wanted to find an answer to the anguished search for self, a search that took them "on the road," following the apparently senseless itineraries of a perpetual vagabondage.

The experience of the writers of the beat generation is so profoundly and typically American that Europeans could glean a better understanding of them through their books than through history or the movies. At the heart of all of their books is the American myth of the frontier and the closely related theme of the journey as metaphor for freedom. It is clearly no coincidence that the greatest political myth of the period would be Kennedy's "New Frontier," which should be attained because it symbolized a new freedom to overtake. It was equally symptomatic that when Kennedy was assassinated writers made him the object of unprecedented campaigns of defamation. America is deeply divided in regard to its own myth: on the one hand, its perpetual restlessness leads it to change and move, on the other, its acquired comfort chains it to the privileges it wants to hold onto.

For Kerouac and the other novelists or poets of the beat generation (from William Burroughs to Allen Ginsberg, from Lawrence Ferlinghetti to Gregory Corso), the life of the vagabond is both hell and the chosen road to scrutiny and knowledge. They travel through space with their own bodies as the vehicle, by means of alcohol, drugs, and sex, continually breaking through new barriers and fences, in order to reach the innermost being, in search of their own identities. Outer space empties into inner space, distance turns into depth, the freedom of the body reflects the freedom of the spirit. But American

Actress Jane Seymour, wearing the "Orénoque" necklace, poses in front of Queen Elisabeth of Belgium's tiara (made by Cartier in 1910) at the exhibition "L'Art de Cartier" held in Paris, 1989.

space, whether geographic, psychosomatic, or even of dreams, no longer sufficed for the wanderings of the beatniks. They were lured by the irresistible seduction of the East. The East is a space that offers itself to the speculation of the soul, the quest for the divine. It is a space in which space no longer matters, where a person could live out his life seated on the same tiny piece of land with crossed legs, because within him "space" is wide open. It was a natural evolution for the beat myth to end at the mysticism of Buddhism or Hinduism, the final and fatal stage of a journey that does not recognize frontiers.

The beat generation defined itself as "beat," that is, both "beaten" and "beatific" in its defeat, liberated, thanks to its marginal existence, from the privileges as well as the prejudices of bourgeois society. In the subsequent decades, thousands of people would reject the consumer society and follow the same route, from the rich West to poor India, from materialism to spiritualism, and, paradoxically, from the unstable to the immovable.

It is not farfetched to see in the experience of the beat generation one of the main sources of myths and dreams for the next generation's uprising in 1968. In that year was the expression of a terrific need for space and movement. Its two most visible and sensational strategies were the occupation of universities (conquest of space) and the organization of enormous marches (conquest of movement). Like the beat generation, the young people of 1968 wanted to fight bourgeois society in the open. They challenged all of its rules: submission to authority, the passive apprenticeship to an arrogant culture, the acceptance of social inequities, an archaic sexual morality. They ridiculed society's contradictions and smashed its idols. These rebels wanted to be outsiders in their own consumer society.

"What do they want?" Instead of responding to its children's revolt, society asked, "What do they want? They already have everything." Faced with a demand that could not be quantified nor understood, the West went looking for hidden interests and motives, especially Eastern: the Chinese, the Kremlin.

Like the representatives of the beat generation, the youth of 1968 became the object of well-orchestrated defamatory campaigns. The West accused them of being dirty (because they wore their hair long) and disgusting (because they enthusiastically embraced sexual promiscuity). These were feeble responses, certainly. The first was paradoxical: in reality, in a Europe that went unwashed for centuries and that for two thousand years had looked on the East with an ambivalent curiosity (attracted by the perfumes and baths, mistrustful of the effeminate manners evidenced by their cosmetics and their leisurely ablutions), it was actually the generation of 1968, which became totally Americanized, that was the first to celebrate the daily ritual of the shower and the union with soap so ardently desired by Europe.

The second argument, relating to their sexual promiscuity, besides being weak, was more interesting. In the first place, amid all its defeats, 1968 carried off a victory, at least partially, on this terrain. It made possible this pleasurable activity in the occupied universities, and in the long term, it led to a sexual liberation of Western civilization that, while still relative, seems unlimited in comparison with the restrictions of earlier times. Moreover, it began the real emancipation of women, only marginally enacted in the 1920s.

On the whole, the cultural confusion of the West that developed after World War II, far from finding any solution, became deeper and more aggravated. The world that arrived at the 1970s was a world where the contradiction between material well-being and psychosocial uneasiness had become explosive. For at least forty years, the various cultural and artistic strains had not con-

Design for Sagittarius, the ninth sign of the zodiac.
Archives Cartier Paris, 1985

verged in any movement of ideas capable of imposing its power on all. On the contrary, the West produced rationalism and irrationalism, figurative and nonfigurative art, rich art and *arte povera*, popular and experimental literature, electronic and pop music, rock and jazz, commercial movies and "auteur" cinema. These are the contradictory expressions of a society pulled between its attachment to tradition and its interest in the new, between its nostalgia for the past and its anxious desire for a new, as yet unexplored, period.

In this transitional phase between the fabulous 1960s and the difficult 1970s, when Europe reached the lowest point of its cultural curve as well as the height of the confrontation between conformity and nonconformity, between the gray-

In this 1983 photograph, Elton John,
a jewelry collector, wears a pair of precious
Les Must de Cartier sunglasses and a contemporary
lapel brooch of platinum and diamonds.

ness of convention and the flamboyance of rebellion, Cartier effected a small revolution of its own. It was small, certainly, in relation to the fate of the world, but it loomed large in the special world of luxury products.

To recap briefly the auspicious story of Les Must, it began in 1968 with the launching of the famous oval cigarette lighter. Five years later, in 1973, its paths and its horizons were clearly drawn as belonging uniquely to the most luxurious lines of production when the first collection of watches in 18-karat gold (named in honor of Louis Cartier) was presented. It was quickly followed by the creation of a new concept of marketing under the new label "Les Must de Cartier," developed by the group that had acquired the company and would be responsible for its direction from that point. Alain Dominique Perrin was in charge of the new label.

The following three initiatives served as the basis for a Cartier renaissance, which set off a new era of splendor. This progress was made possible by the convergence of several serendipitous intuitions.

In the first place, the intense process of the redistribution of revenues that had been going on for at least a quarter of a century resulted in a relative social leveling. Comfortable standards of living were enjoyed by more and were more widely spread rather than concentrated; that is, the base of the pyramid of wealth had been enlarged and its height reduced. For the companies selling luxury products, it meant facilitating access to luxury, reaching an expanded clientele with products that, while retaining an aspect of preciousness and elitism, no longer or rarely carried prohibitive prices.

In the second place, a certain "conquest of space" had to be undertaken. The process of distribution essentially based on the three historic shops in Paris, London, and New York had become outmoded. These legendary locations, especially the first, would remain the centers of reference and distribution, the gravitational poles of attraction, but around them would revolve the more mobile satellites of the boutiques and various distributors. In this way, Cartier would become accessible to a greater public, geographically (in physical proximity) and socially.

In the third place, the concept of beauty for beauty's sake, like that of art for art's sake, was no longer dominant, as it did not correspond to the pragmatism of the time. Beyond the fact that only a very limited number of clients could afford fabulous parures, they were also increasingly frowned upon as ornamentation that serves no purpose. A transformation of taste was at work: from interest in a beautiful object that served no purpose to an object that was beautiful because it was useful—that is, from jewels to functional accessories. Cartier knew how to take advantage of this change in taste without slowing down its most revolutionary aspect. It began with watches but did not concentrate on them exclusively; in fact, after the cigarette lighter, the range of Les Must soon encompassed, in the space of a few years, leather goods, pens, the "arts of the table," perfumes, and glasses. In obeying the necessity to diversify its production, the inevitable corollary to the enlargement of its system of distribution and its clientele, Cartier kept its essential nature. From the beginning, it simply added to its production of jewelry that of watches and luxury accessories. In the case of certain types of objects, such as pens, handbags, travel accessories, and lighters, it had a legacy as rich and varied to draw on as it did for jewelry and clocks and watches.

These three overarching developments that presided over the reestablishment of Cartier's name and its spectacular renaissance came to be linked with a fourth that embraced them all: in a period in which the desire for beautiful, rare, and precious objects was focused more and more on objects from the past, the eclecticism that Cartier had already exploited for several decades opened up fascinating possibilities for development in the immediate future. As the decorative arts had seen only a modest evolution since the 1930s, the combination of Cartier's creations of the past were simultaneously classic and contemporary, traditional and new, mature and modern. While these creations give the impression of gliding from the Garland style to Art Deco, from "tutti frutti" to the naturalism of flora and fauna, from geometric abstraction and Western rationalism to the sensuality of volumes and colors of the East, it is through their alternation and combination that these various styles over the decades defined a "Cartier style" that from now on would make for a homogeneous universe.

Leafing through the illustrations of Cartier's most recent production gives a sense of the freedom with which the jeweler, since the 1970s, has revisited and realized its past while bringing it up to date by means of a fortunate creative eclecticism. Its inspiration has at its command limitless themes, from rigorous geometric compositions in diamonds on platinum to the chromatic explosion produced by the combination of three colored gems (emeralds, sapphires, and rubies), or to the inexhaustible proliferation of the animal theme, from the variety of lines and volumes to that of the many ways to shape the stones.

Platinum's role in this highly mobile and untiring "revolution of Cartier around its axis" is not fixed. Setting aside for the moment the area of watchmaking, it is necessary to distinguish in its jewelry production two clearly differentiated lines: traditional fine jewelry and New Jewelry, a radical innovation, introduced in the 1980s by Micheline Kanouï (Cartier's Creative Director for fine jewelry) and intended to evolve in phases (each new collection is inspired by a unifying "monographic" theme). While platinum continues to play

its role as the ideal, and thus irreplaceable, setting for diamonds in fine jewelry (as well as the other gems, particularly sapphires), in the domain of the New Jewelry and its jewels, which are less exclusive, more affordable, and made as production rather than unique pieces, gold clearly allows the greatest freedom in terms of style.

The single solid victory achieved by 1968, the equality of men and women, was fully confirmed

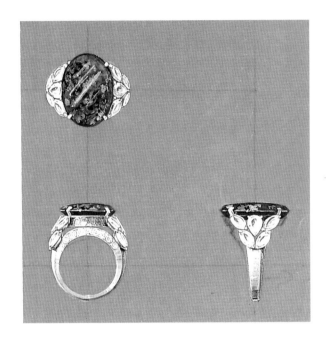

Design for the "Anastasia" ring.
Archives Cartier Paris, 1987

in the 1980s. Sporty, dynamic, professionally active, women embraced an entirely new simplicity when it came to their behavior and their clothes. In the realm of fashion, freedom of the body was paramount, and the desire for mobility dictated. The unisex concept, which never affected more than a narrow segment of youth in the 1970s, had grown outdated, although jeans, symbol of youth's carefree attitude, had broader social acceptance. The aristocratic and somewhat formal coldness of platinum did not suit the woman who tended to present an image at once spontaneous and convivial, sensual and sunny, exuberant and optimistic.

hard-working and dynamic. Jewels had to come out of their jewel boxes and become, like the woman of the 1980s, young, active, mobile, creative. This shifted the emphasis to gold, a metal that presents a warmer, softer, younger image than platinum, easier to work, more casual in terms of occasions on which it can be worn, lending itself to the growing tendency to disregard the distinction, save on exceptional occasions, between day and night.

However, platinum continued to dominate a sector that has played a particularly interesting role in Cartier's recent history: animal jewelry, on which Cartier's contemporary designers have concentrated their creativity. The great success of this theme after World War II did not diminish; in fact, it did not reach its full maturity until the 1980s. The panther, which became Cartier's living symbol, with its aggressive sensuality, agility, and its egocentric yet angelic beauty, could also be said to be the perfect emblem of the contemporary woman. It is hardly surprising that Cartier would study its expressive potential not only from a plastic point of view but also with a penetrating psychology: its movements and its poses, its silken but menacing poise, ready to spring at any moment, its smooth yet deadly leaps, its yawns and roars. These were turned into jewels, watches, and accessories, some fashioned with a highly detailed naturalistic iconography, others with an almost abstract stylization. For all these variations on the cat and other creatures (Cartier gave birth to a large bestiary, a sort of precious Noah's ark, which forms the most original aspect of its contemporary production), platinum was generally preferred to gold, given the fundamental role of diamonds and, among the colored gems, especially emeralds and sapphires in those pieces representing the feline coat, birds' plumage, or the eyes of any animal.

Watches merit a separate discussion, as they

have occupied a central role in the jeweler's universe since the first collection, dedicated to Louis Cartier, was launched. They also illuminate style

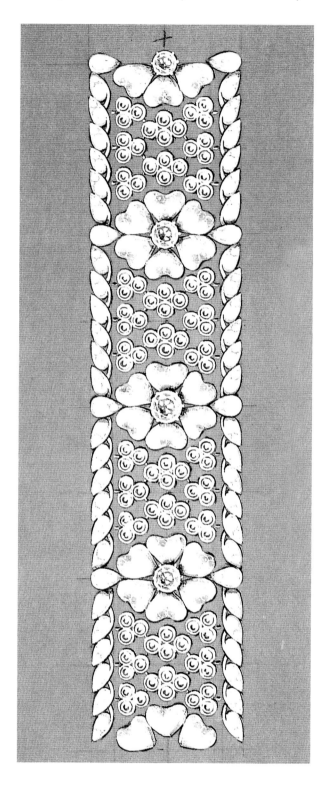

Design for the "Mayerling" bracelet.
Archives Cartier Paris, 1988

matters and the fluctuating fortune of platinum in particularly interesting ways.

The tremendous success, remarkable in Louis Cartier's day but extraordinary from 1973 on, that the jeweler enjoyed in the area of watches owes much to the purified, geometric, and timeless lines of the wristwatches. Realizing that today's Tank watches drew on the first model dating from 1917 for the design's basic form is amazing. In an object as basically simple as a watch, in which technological progress need not dictate the outer sheath of the mechanisms, it is astonishing to note that Cartier reached the optimum balance between aesthetics and function from the beginning of the century and had maintained it up to the present. The only technical innovation that had any effect on the watch's appearance was the progressive shrinking of the movements, which allowed the dials, especially in women's watches, to be enlarged (which made them easier to read) without having to make the cases heavier. This development was much less marked, or much less noticeable, in the case of watches for men. As the Cartier collections continue to demonstrate, whatever movement is used, mechanical or quartz, it will work perfectly in an Art Deco case while handily—and magnificently—satisfying all the technical and functional, not to mention aesthetic, requirements.

As was noted earlier, in the chapter on Art Deco, platinum's beauty when presented alone and its ability to embody with an incomparable discretion and an expressiveness natural to polished metal the character of the elegant machine that the wristwatch presents was admirably exploited by Cartier since the 1920s. Throughout the era of Les Must, versions in platinum were offered for the most important models (and, for women's watches, in platinum set with diamonds). Nonetheless, in the 1980s platinum was rejected in favor of gold even more decisively in watches than in jewelry. In

order to understand this phenomenon, and its converse, that is, the return of platinum in the 1990s, it is necessary to realize that the impact of platinum by itself is very different from that of the metal covered with gems. The association of platinum and diamonds in jewelry goes back to a tradition, since then well established, of aristocratic refinement and constitutes the ne plus ultra of luxury. In watchmaking, plain polished platinum does not show off its precious value and even seems, with a calculated insouciance, to take pleasure in passing itself off as an ordinary metal, which made it unattractive in a period characterized by a pronounced tendency to display rather than to hide wealth.

The crisis of cultural ideals and values that is the keynote of postwar Western civilization reached its apex in the 1980s, at the end of a serious economic recession that had shaken the foundations of the capitalist system in the second half of the 1970s. The oil crisis of 1973 created shock waves that affected the economy for the next five years, setting off in all the Western countries a spiral of almost uncontrollable inflation accompanied by a prolonged stagnation—to the great surprise of the economists, who had claimed such a combination heretofore impossible. When this condition began to turn around at the end of the 1970s and the first signs of a strong recovery were noted in the West, the psychological effect was comparable to that produced by the end of the war. It set off a period of widespread optimism and a race for pleasure.

The 1980s were a period of carefree hedonism. This pendulum movement and its cyclic nature are well known—periods of hardship give way to periods of well-being, while the social cohesion (more or less forced) of the former dissolves into a pronounced individualism in the latter. In other words, humans might by nature tend to share their misfortunes with their neighbors, while they savor their good fortune within an intimate domestic circle. And the prosperous 1980s, corresponded to a period of individualism, not to say egoism; the reevaluation of that word, customarily laden with negative connotations, is the most important contribution of the yuppies to Western cultural history. The absence of moral barriers to the new bourgeoisie's season of pleasures is typical of the 1980s. Comfort and wealth no longer needed to be hidden or disguised; on the contrary, they were something to display, like a smile or a good tan.

Cartier, strong in a tradition that guarded against excess, declined to give in on the issue of taste and maintained its production in the confines of a "contemporary classicism," which was its inimitable keynote in terms of style. From time to time, when it sensed a countervailing wind blowing, it offered platinum again. It also anticipated currents to come: from the reissue of the famous Rolling Rings in the all-platinum version (page 228) to the "CC" cuff bracelet (page 229), from the collection of Louis Cartier watches in platinum (with burgundy Roman numerals on the dial, pages 226–27) to some models of the New Jewelry. Even in the last area, where gold's predominance was especially strong, platinum replaced gold from time to time, sometimes in the same model, either to satisfy a client's special request (the fashion for platinum invaded Japan at the very moment that its use was greatly reduced in Europe) or to enlarge and diversify the collection by offering even more selective new versions, in limited and numbered editions.

The fact is, however, that in the 1980s the time was never propitious to offer restrained platinum as a large-scale alternative to gold, which was not only warmer, it also undeniably made a bigger impact.

The 1990s, which were immediately preceded by an abrupt and alarming economic downturn, and which introduced the historic turmoil of the fall of eastern Europe's Communist regimes, immediately made clear its intentions to reexamine the individual hedonism of the previous decade. The total absence of positive myths in the West, cultural as well as political, persisted, and a certain tendency to tone down signs of wealth returned, effectively a social slowdown. The 1980s had been a period of frenetic mobility, marked by very rapid social elevation. The 1990s brought a better balance and consolidation of positions. It is clear that if the nouveaux riches are more easily tempted by ostentation, the pleasure of the understatement would reveal itself in the fullness of time.

Meanwhile, the economic and social rise of women and the increasingly determined way they affirmed their equality with men provoked some interesting repercussions, among them a new focus on men's fashions. Between the two sexes, an exchange of values comparable to the exchange of warmth between two bodies of different temperatures when they make contact seems to have taken place: men have lowered their socioeconomic temperatures and raised their physical and aesthetic temperatures. If men were no longer the only ones to assume a commanding role and to carry the reins of power, women were no longer alone in embodying beauty, taking care of their bodies, surrounding themselves with an aura of enticement. While waiting to learn the art of seduction, Western men have thoroughly mastered all the movements and poses of narcissistic vanity. This process, which obviously goes against cultural prejudices, has just begun and clearly has some distance to travel. However, the more tolerant attitude toward the relative masculinization of women and the corresponding feminization of men has already brought about certain shifts in the domain of clothing and, our own particular interest, accessories. Although men still do not wear actual jewels, they are inclined to think of the precious, ornamental accessories they use as such—cuff links, lighter, key ring, and, especially, watch. Men see the watch less and less as a functional object intended to measure time and more and more as a decorative object intended to reflect the personality of the wearer. Platinum, with its precious nature and rarity, serves this purpose well. On the one hand, it assures discretion (a value that continues to be asserted in men's clothing the further up one goes on the social ladder), and, on the other, it carries the aura of technical research and beauty. Cartier, in line with the "aristocratic" image of this metal it always cultivated, tends to use platinum to clothe the most sophisticated mechanisms, the rarest and the most expensive models. Obtaining enough platinum for the case of a single watch requires the extraction of about ten tons of platinous rock, from a depth of 3,000 to 5,000 meters—certainly a suitably precious coat for a refined mechanism of sophisticated technique. Thus, the restrained and discreet beauty of platinum is used to enclose in its heart an even more secret beauty.

Today, platinum is rapidly regaining the position it had lost in jewelry—not only in fine jewelry, and in the many variations offered on the animal theme, but also in a more everyday area: engagement and wedding rings (page 257), emblematic jewels for which Cartier audaciously uses platinum, symbol of precious nature, endurance, of marriage.

We have seen how platinum, faithfully accompanying the transformations of style and society, in a century changed its language and its destiny: from the select metal intended for creations of unmistakably exceptional character (due as much to the social position of those wearing it as to the occasions that led them to wear it), it has become progressively the select metal intended for objects that are always worn, through day and night, and worn always, throughout one's days,

Design for the "Orchidée" earrings.
Archives Cartier Paris, 1990

like watches and engagement and wedding rings. By virtue of the route it took from the extraordinary to the everyday, the most precious metal embodies and sums up with its accustomed strength the route of our century: from fables to reality, from dreams to lucidity, from Picasso to the post-modern, from Stravinsky to Prince, from the operatic "malady of existence" of Thomas Mann's *The Magic Mountain* or Italo Svevo's *The Confessions of Zeno* to the less threatening abscesses of the minimalists. But also, not to deny it, from penury to abundance, from privileges to balance, and from the extraordinary to the everyday. The exceptional on the finger or the wrist, day after day, at all hours. The man and woman of the year 2000 could say, to paraphrase Flaubert, "The Belle Otero, c'est moi."

Design for the "Bantou" bracelet.
Archives Cartier Paris, 1992

THE FUTURE OF PLATINUM

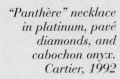

"Panthère" necklace in platinum, pavé diamonds, and cabochon onyx. Cartier, 1992

*I*n jewelry, platinum enjoyed a period of extraordinary splendor between the beginning of the century and World War II, when it served as the great interpreter of Art Deco. In 1939–40, because of its multiple industrial uses and its strategic importance, platinum was requisitioned for military use in countries at war and thus it completely disappeared from the market.

With the postwar economic recovery, the profound social changes brought about by the conflict did not fail to affect the market for luxury products, particularly jewelry. For a great many reasons to do with taste, fashion, various decorative styles, design, economic well-being for a larger share of the population accompanied by a narrower concentration of wealth, until recently platinum has been used almost exclusively in fine jewelry—that is, in rare and costly pieces that demand a platinum setting for diamonds. In addition, since the market for fine jewelry had become extremely restricted and limited, the amount of platinum used has gradually declined over several decades.

Only two markets in all the industrial countries depart from this trend: in Japan, where platinum has traditionally been preferred to gold because its color works better with the Oriental complexion, and in the Muslim countries, where it is highly sought after for an entirely different reason. The Islamic religion forbids men (but not women) to wear gold, while platinum is allowed for the simple reason that its existence was entirely unknown when Muhammad set down this rule.

Today, however, platinum has made a strong comeback on the world scene. The very reason it lost popularity in the ostentatious 1980s—due to its understated appearance—has made it highly desirable in the 1990s, where it is now appreciated for being rarer and more precious than gold as well as more discreet. As the 1980s ended and the 1990s began, platinum proved to be the perfect interpreter of the spirit of the time and the values of elegance, discretion, and seriousness that seemed to accompany humanity in its move toward a new millennium. The acquisition of a platinum jewel, for oneself or as a gift to someone precious, strongly indicates a clear and deliberate choice that reflectes the wearer's personality as well as a quest for the rare and the perfect.

The most distinguished jewelers today select platinum to set the distinctive tone of their most refined collections. The material's intrinsic qualities—purity, incorruptibility, luminosity, color—make it a particularly appropriate way to express eternal love and the enduring marriage; it is no coincidence that engagement and wedding rings have led off platinum's return to jewelry, not only in Japan and the Near East but also in the major Western countries.

Smooth or satin-finish, a perfect white alone or combined with the warm color of gold, with or without gems, platinum submits to the jeweler's skills with extraordinary ductility and malleability, offering a remarkable range of creative possibilities.

Its combination of restraint and strength has recently made it the metal preferred by men, who have increasingly chosen it for luxurious accessories, especially watches. All of the finest watchmakers have been providing their most important models with platinum cases and their most technically complicated models with platinum movements.

Since 1990, the use of platinum in jewelry has seen a meteoric rise, and it holds out the promise of interesting future developments. Today, jewelry makes use of forty percent of the world production of platinum. A worldwide trend toward a more sophisticated taste in jewelry, led by increasingly experienced and demanding clients, has accompanied the quest for a more positive and intimate life-style. Platinum, with its unparalleled combination of discretion and refinement, should easily satisfy such demands.

"Mini-Panthère" bracelet watch in platinum and diamonds. Cartier, 1993

70

Louis Cartier (1875–1942) portrayed at the age of twenty-nine by Émile Friant. It was he who made the decision to move Cartier's main store to the rue de la Paix in 1899. In the same period, he introduced the art of platinum to Cartier.

THE GARLAND STYLE
From Its Origins to 1914

After exciting the interest of scientists, platinum won over jewelers. Cartier, who wove it in very fine threads to enhance the brilliance of diamonds, contributed to its legitimacy as an irreplaceable precious metal. As the light of the Belle Époque died away, it set off sparks that would become legendary.

The Eiffel Tower and the Champ-de-Mars in Paris, 1889
The construction of the Eiffel Tower in metal (designed by Gustave Eiffel for the World's Fair of 1889) became the symbol of an international movement, of a capital (Paris), and of a society in profound transformation.

1865

Date			N°	Nom		Description						
Décembre	4		12	Occasion	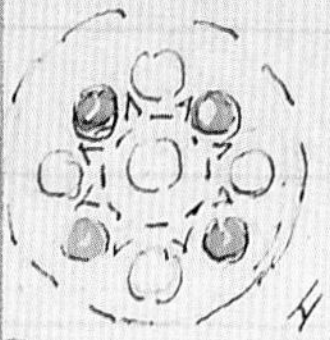	Plat or mat 5 mi perles	"	"	81			12
"	"	"	"			4 émeraudes filets email noir	50	"	Octobre	6		"
"	"		13	Savart	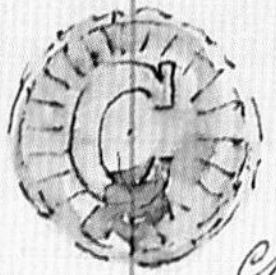	Plat or mat rayon	"	"	82			13
"	"	"	"			Chiffre 4 C. T. 4. Or 3 gr 70	14	75	Juillet	1		"
1866 9bre	9		14	Delavigne	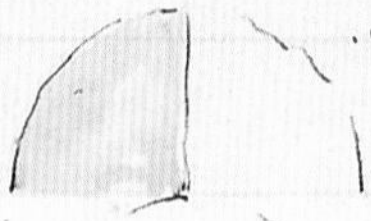	Bombé moitié	"	"				
"	"	"	"			or mat et platine	80	"				
1867 Février	2		15	Thomas		Or rouge et	"	"	70 Juin	13		15
"	"	"	"			mat têts de	"	"	"	"		"
"	"	"	"			chien argentcivile	52	"	"	"		"
" Avril	9		16	Mekay		Or poli guilloché email	"	"	71 xbre	9		16
"	"	"	"			noir filets	27	50	"	"		"
" Mai	31		17	Occasion		Double or mat etoiles	"	"	72 xbre	30		17
"	"	"	"			poli mi perle	60	"	"	"		"
Septbre	11		18	Thomas		Argent oxidé bordpoli	"	"				
"	"	"	"			chiffre A. B. or rouge	18	"				
Octobre	11		19	Lièvin	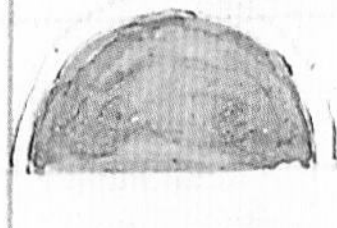	Lapis plat encerclé	"	"	70 Juin	27		19
"	"	"	"			monture genre poli	55	"	"	"		"
"	"		20			Rond onyx encerclé	"	"	71 Aout	19		20
"	"	"	"			filet or 1 mi perle	44	"	"	"		"
"	"		21	Davie et Dumand		Rond bombé quadrillé	"	"	73 Mai	21		21
"	"	"	"			filigrane or mat	32	"	"	"		"
"	25		22	Codan		Rond plat poli			70 Avril	24		22
"	"	"	"			guilloché poix	33	"	"	"		"

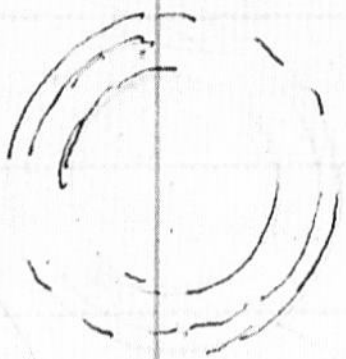

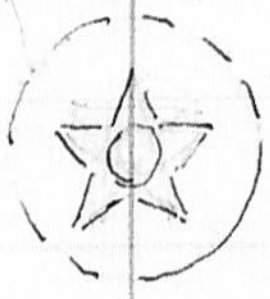

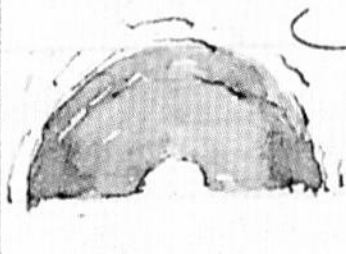

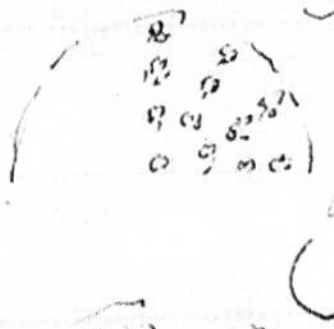

Opposite: *Part of a page from a book in which Cartier kept track of its inventory in 1865. Cartier's artistic heritage, an inexhaustible source* "matte gold bracelet, [with] openwork ornaments on a platinum base," *and in 1899 to a* "hat pin, [with] two-sided ornament, rose[-cut diamonds]

that Cartier has mined to recover its most fruitful creations, is comprised of nearly 350,000 archival documents that cover almost 150 years of a multi-faceted history. Drawings, sketchbooks, order books, inventory books, more than 30,000 glass plate negatives, photograph albums, and client correspondence have been conserved in Paris, London, and New York, the three historic branches of the Maison Cartier. The page opposite, one of many decorated with quick sketches, contains a drawing of one of the earliest objects to use platinum: a pair of simple cuff links with convex disks, half matte gold, half platinum, recorded on 9 November 1866.

In the second half of the nineteenth century, platinum was frequently combined with matte gold. In 1882, for example, there is a reference to a

mounted in platinum." This last is of particular interest, as it concerns the earliest head ornament in platinum; tiaras did not appear until after 1900. A study of the inventory books reveals that starting in 1899, the year in which Cartier moved to 13, rue de la Paix, Cartier's golden rule becomes platinum.

Above: *Design for a* plaque de cou *(the clasp of a dog collar that can be worn as a brooch), pencil and ink on tracing paper, 2½ x 6" (6.7 x 15 cm). Archives Cartier Paris, ca. 1905*

The ink drawings in Cartier's archives demonstrate the careful attention to detail, including suggestions of volumes and hollows through subtle shading, with which the jewelry designers communicated their ideas to the artisans who executed them.

Left: *Design for a tiara in platinum and diamonds. Archives Cartier Paris, ca. 1902*

Opposite below: *Pendant watch, front and back, in platinum, gold, white enamel, diamonds, and porcelain. Cartier Paris, ca. 1900. Cartier Collection*
The Roman-style subject in Wedgwood porcelain is framed by a border of white enamel. Platinum serves as a decorative element in the form of a border and a watch ring with rose-cut diamonds in a millegrain setting.

Below: *Escutcheon pendant in platinum and Wedgwood porcelain. Cartier Paris, 1905. Cartier Collection*
This pendant was set entirely in platinum. The two tassels are articulated.

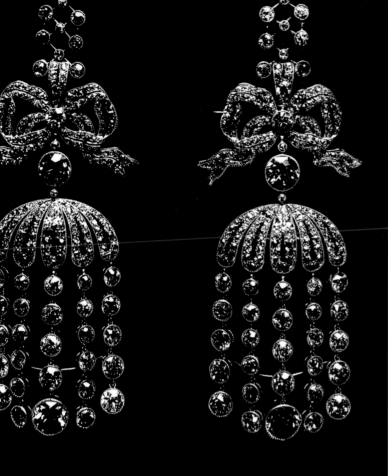

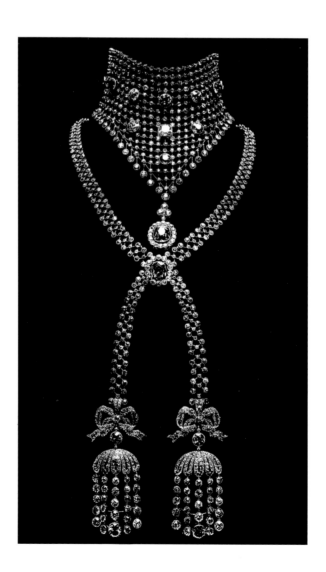

Résille *necklace in platinum and diamonds.*
Archives Cartier Paris, 1903
*This extravagant piece made effective use of plati-
num's special qualities. The settings are so thin
they almost disappear, enhancing the fire of the
diamond collets. The articulated links give a
flowing look to the collar with pendant and make
it as easy to wear as a piece of fabric. Only platinum could permit a setting so thin yet with the
strength to support the considerable weight of the
two tassels.*

*This necklace, inspired by Marie-Antoinette's
famous necklace with tasseled ribbons, was a special order from Spanish dancer Caroline Otero,
known as La Belle Otero, and utilized the stones
from her famous bolero, created by Paul Hamelin.*

*T*iara in platinum and diamonds. Cartier Paris, 1905. Cartier Collection

This tiara, crowned by seven pear-shaped diamonds, is only one of many such creations made in the Garland style, intended for the crowned heads and aristocracy of Europe and other lands as well. In the space of a few years between the beginning of the century and the onset of World War I, Cartier became the official purveyor to the courts of Portugal, Russia, Serbia, Italy, Belgium, and Egypt.

*T*wo geometric brooches in platinum and diamonds, respectively dated 1906 and 1907. Cartier
Collection
The brooch on the left is hexagonal, with a central
sunburst pointing its rays toward the laurel-leaf
border. The lozenge-shaped brooch on the right
has nine diamonds crowned with double arches.
The thinness of the platinum setting emphasizes
the play of volumes and the contrast between filled
and empty space.

At the beginning of the century, couturiers and jewelers worked together closely. The collaboration between Charles Worth and Cartier was further strengthened by the bonds of matrimony.

Platinum facilitated such a thin setting for diamonds that they appeared to be sewn on the fabric seen between the metallic support and the stones. In addition, this "fabric" could reproduce the motif of the dress, giving the illusion that the diamond motifs were integrated with the dress.

Above: *Dog collar in platinum, diamonds, and velvet. Cartier Paris, ca. 1905. Cartier Collection*
Lines, rosettes, and interlacing figures in diamonds alternate on a ground of midnight blue velvet.

Below: *Bow brooch in platinum, diamonds, and black fabric. Cartier Paris, ca. 1906*
The feather was another popular subject for brooches at the beginning of the century. While goldsmiths who favored the Art Nouveau style made them in such materials as translucent enamel, Cartier the jeweler rendered the subject in platinum set with pavé diamonds and endowed it with

a lightness that called the real object to mind. These brooches were sometimes provided with a setting that created a "trembling" effect, known as a tremblant setting.

Opposite left: *Feather brooch in platinum and diamonds. Archives Cartier, ca. 1905–6*
From the spine of the feather alternating rows of closely set diamonds and diamond collets are projected, with a cushion-cut diamond at the lower end in the knot of a triple bow. The stones are all mounted in a millegrain setting.

Opposite right: *Feather brooch in platinum and diamonds. Archives Cartier Paris, 1914*
The lines set with diamonds curve in to create a scroll-like effect. The brooch is "tied" at the lower end with an ornamental bow.

Records specify that this jewel was created on special order. Since the prevailing style of Cartier's stock in 1914 was already Art Deco, it appears that this client specifically requested a Garland style jewel.

As pendant brooches grew longer they grew more elegant, but because of their platinum settings they remained light and flexible, and thus easy to wear.

Tassel pendant brooch in platinum, diamonds, and velvet. Archives Cartier Paris, 1907
The tassel hangs from two diamond rectangles on black velvet. This part could be detached from the convex triangular brooch, which is decorated with a brilliant-cut diamond in its center and divided in three by a triple row of diamonds.

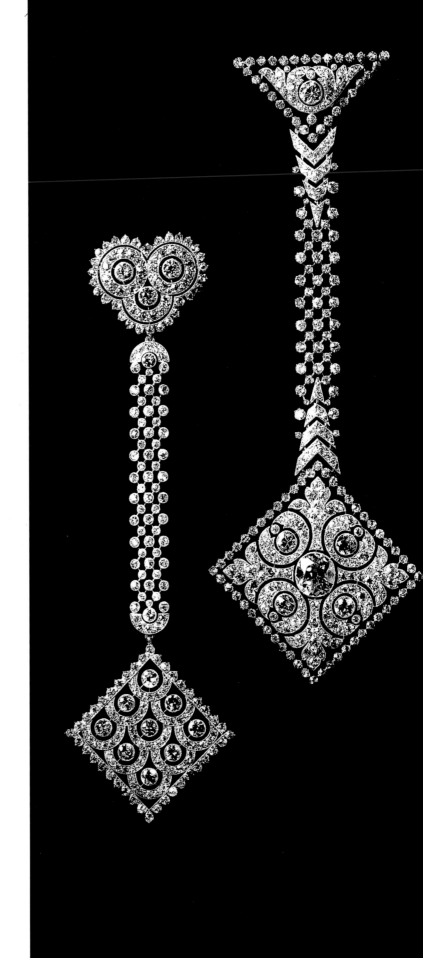

Near right: *Pendant brooch in platinum and diamonds.*
Archives Cartier Paris, 1909
The lozenge-shaped brooch decorated in a fish-scale motif hangs from a chain of diamond collets attached to a detachable clover brooch.

Far right: *Pendant brooch in platinum and diamonds. Archives Cartier Paris, 1909*
The lozenge-shaped brooch hangs from a chain of diamond collets with pavé diamonds forming arrows at either end. It is attached to a triangular brooch.

*T*he bow has always been one of the favorite ornaments of goldsmiths. After platinum was introduced, the goldsmith-jeweler could attain an unparalleled refinement with the shape, especially by means of bringing out the sparkle of the stones.

Right: *Bow brooch in platinum and diamonds. Cartier Paris, 1907. Cartier Collection*
The double bow is set in pavé diamonds, with a round diamond in the center of a lozenge-shaped openwork motif. Each of the three drops is composed of a line of diamonds of increasing size.

Near left: *Bow brooch in platinum and diamonds. Cartier Paris, 1906. Cartier Collection*
The triple bow is set in pavé and navette-cut diamonds "tied" with a cushion-cut diamond knot surrounded by a small crown of diamonds. The three-dimensional lace motif is detachable.

Opposite left: *Brooch in platinum, diamonds, and sapphires. Archives Cartier Paris, 1907*
This brooch, set with rose-cut and brilliant-cut diamonds, has a central lozenge-shaped motif, with diamonds and two crossing lines of calibré-cut sapphires, surrounded by fluttering ribbons. Two free-hanging drops of interwoven diamonds and square-cut sapphires hold two cushion-cut sapphires framed by a double row of diamonds.

Opposite right: *Epaulet brooch in platinum, diamonds, and pearls. Archives Cartier Paris, 1907*
The top of the jewel, consisting of a button pearl surrounded by diamonds with a triple chain of diamonds, is connected by a diamond motif to a central lozenge with a button pearl surrounded by diamonds in a laurel-leaf pattern and crowned with a double bow. A third button pearl is suspended from the lozenge, set inside a pear-shaped motif with laurel-leaf pattern and flanked by two aiguillettes in millegrain-set pavé diamonds. It was made on special order for Princesse Marie Bonaparte.

One of the most popular jewels of the Belle Époque was the corsage brooch, or stomacher, which could attain surprising dimensions, in some cases forming a garland that reaches from one shoulder to the other. A great many of these jewels could be mounted on a platinum frame, and the entire flexible brooch locked together to become a tiara.

Above: *Corsage brooch in platinum and diamonds. Archives Cartier Paris, 1910*
The acanthus-leaf and scroll decoration consists of brilliant-cut and rose-cut diamonds highlighted by several cushion-cut diamonds. The upper border is composed of a line of collet diamonds. A central pear-shaped motif is suspended below, flanked by two tassels.

Above: *Corsage brooch in platinum and diamonds. Cartier Paris, 1912. Cartier Collection*
This large Garland style brooch consists of three swags and a central floral motif. Four hanging laurel-leaf pendants end in a diamond. A pear-shaped diamond framed by small diamonds hangs from two small diamond "chains" below.

Left: *Corsage brooch in platinum, diamonds, and pearls. Cartier Paris, ca. 1907. Cartier Collection*
This jewel takes the form of a leaf-decorated triangle with incurved sides. At its center, a rosette is embellished with five pearls. Two floral-decorated drops hang on each side, each ending in a pearl. A pear-shaped diamond framed by small diamonds is suspended below.

*M*onograms and magic numbers, or good-luck charms in the form of round brooches or buttons of any shape are perfectly readable because platinum can be employed in extremely thin settings.

Opposite: *Examples of monograms in platinum and diamonds. Archives Cartier Paris, 1906–9*
These monograms are openwork brooches or buttons, with intertwined letters, set with brilliant-cut or rose-cut diamonds.

Right: *Sautoir in platinum, pearls, and diamonds. Archives Cartier Paris, 1909*
The pearls are tightly lined up in nine rows. The diamond bars flank a round central motif that carries two intertwined Ps surmounted by the crown of the czars. It was made on special order for Grand Duke Paul of Russia.

Below: *Monogrammed buttons in platinum and rose-cut diamonds. Archives Cartier Paris, ca. 1905*

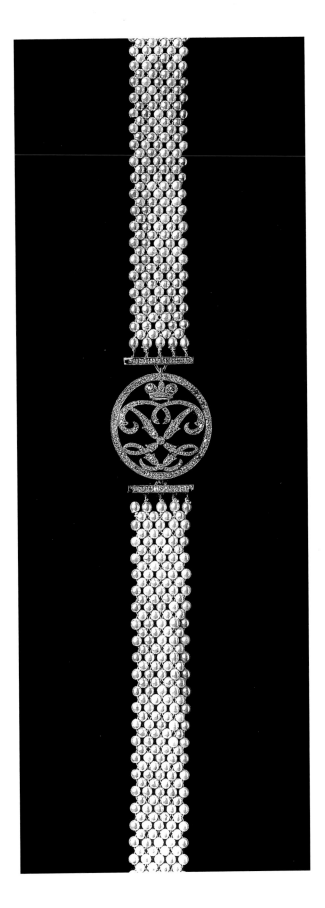

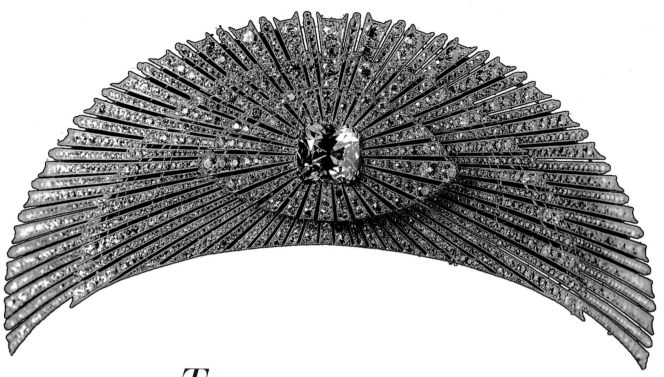

*T*iaras, diadems, bandeaux, and aigrettes were the head ornaments at the time that the Garland style was blossoming at Cartier. They encompassed a wide variety of subjects and ornamentation.

Above: *Kokoshnik-style sunburst tiara in platinum and diamonds. Archives Cartier Paris, 1908*
In this Russian-style tiara, rows of diamonds form rays around a navette-shaped motif in relief set with a cushion-cut diamond of 17 carats.

Opposite above: *Bandeau in platinum and diamonds. Archives Cartier Paris, 1913*
This jewel inspired by the East has a flexible base set with pavé diamonds from which a triangular central motif arises, decorated with openwork arabesques and topped by a stylized feather with eleven curving elements set with rose-cut diamonds.

Opposite below: *Rectangular sunburst plaque de cou in platinum and diamonds. Archives Cartier Paris, 1909*
A cushion-cut diamond shoots its rays past the oval motif to stop short of the double frame consisting of two lines of diamonds.

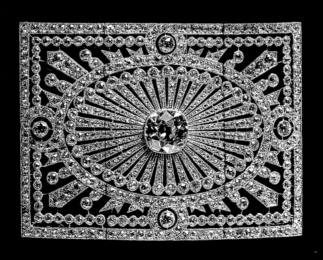

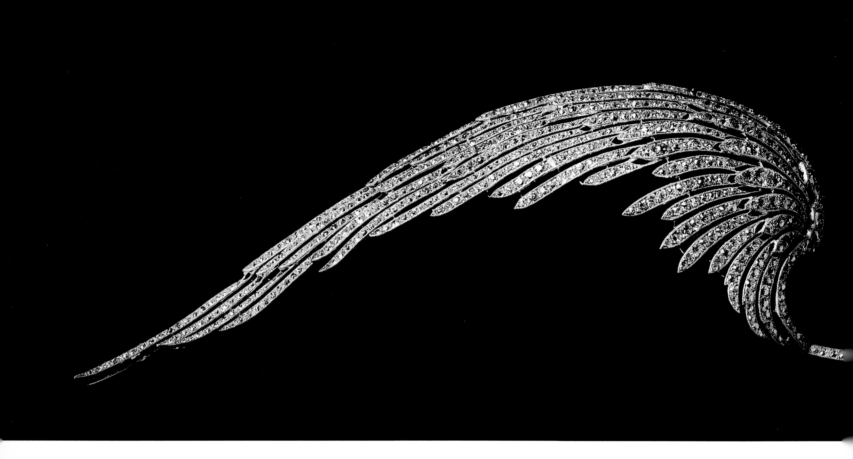

*L*ike the rays of the sunburst motif, the feathers of the wing motif are composed of graduated and overlaying rows of diamonds, wonderfully animated in order to create elegant movements that suggest symbolic flight.

Above: *Wing tiara in platinum and diamonds. Archives Cartier Paris, 1910*
This tiara was made on special order for Lady Ross.

Opposite: *Corsage brooch in platinum, diamonds, and pearls. Archives Cartier Paris, 1906*
This is a stylized version of the motif of wings that support a group of free-hanging garlands in floral and bow decoration. A central pendant with pearls is flanked by two rows of diamonds of increasing size.

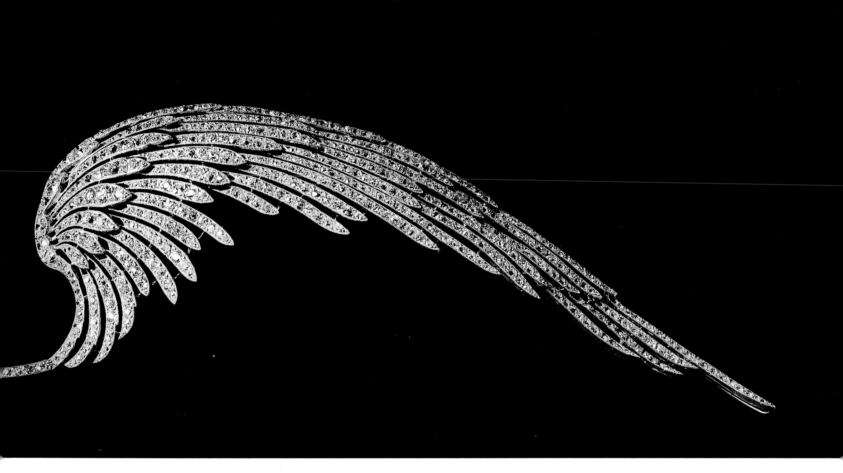

*D*uring the first ten years of the twentieth century, many jewels took on staggering dimensions, especially the corsage brooches and pendant necklaces, all in diamonds or diamonds and pearls. Subsequently considered impossible to wear, such creations were drastically altered, which is why so few of them can be found today in their original state.

Opposite: *Corsage brooch in platinum, diamonds, and pearls. Archives Cartier Paris, 1910*
A sash of pearls connects two very different diamond decorations. A spiral rosette with a central brilliant-cut diamond supports two tassels, each suspending three pear-shaped diamonds, and five pendants composed of pearls and diamond motifs. Four rows of pearls attach this decoration to another in platinum and diamonds in the form of a cap held by a bow, suspended in its turn by an openwork lozenge-shaped diamond brooch.

Right: *Pendant necklace in platinum and diamonds. Archives Cartier Paris, 1910*
The chain of the necklace is composed of ten elongated plaques with hexagon-cut diamonds at their center linked by diamond lily of the valley motifs and a navette-cut diamond. Suspended from the round element featuring a brilliant-cut diamond in its center is a hexagonal openwork pendant decorated with a geometric Islamic-style design including seven free-hanging hexagon-cut diamonds with a total weight of 27 carats.

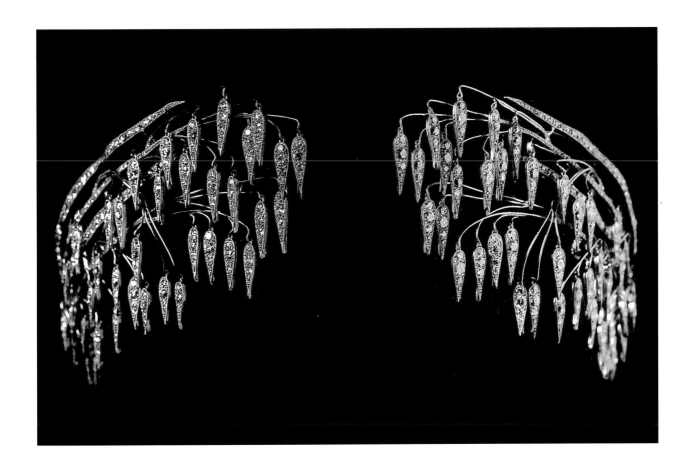

Flowers and other plants furnished additional themes for tiara ornaments.

Above: *Wheat-ear tiara in platinum and diamonds. Archives Cartier Paris, 1911*
The stylized stalks of wheat are made up of graduated diamonds set with the bottom pointed down. The ears move freely, suspended by a great number of extremely fine platinum threads. Two stalks of wheat reach toward each other but do not meet in the center of the tiara. The tiara can be converted into a corsage brooch.

Opposite: *Lily tiara in platinum and diamonds. Cartier Paris, 1906. Cartier Collection*
This tiara, which could be converted into a corsage brooch or a necklace, is composed of two sprays of lilies set with old mine-cut and rose-cut diamonds. A large round diamond marks the intersection of the two sprays. The necklace originally belonged to Mathilde Townsend Welles, wife of Sumner Welles, undersecretary of state to President Franklin D. Roosevelt. Its last owner, Thora Ronalds McElroy, inherited it. Cartier bought it back at an auction held by Sotheby's, New York, on 23 April 1991.

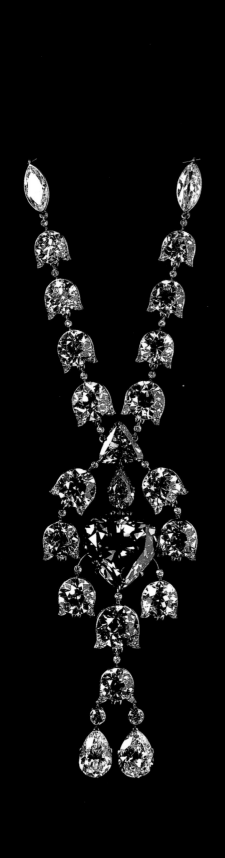

The lily of the valley motif was formed by a delicate bell of platinum in which five diamonds, one large and four smaller ones, were tightly set, giving the impression of a single stone.

Left: Corsage brooch in platinum, diamonds, and blue diamonds. Archives Cartier Paris, 1910
This jewel features a heart-shaped blue diamond of 30.82 carats. It is crowned by a pear-shaped diamond and a triangular blue diamond. These three stones are held by two rows of diamonds set as lilies of the valley that give way to two navette-cut diamonds and are framed by eight more diamonds set as lilies of the valley. The central element ends in two diamonds suspending two pear-shaped diamonds. The jewel was sold to Mrs. Unzue.

The heart-shaped blue diamond (often referred to in reference works as the Unzue diamond) was sold in 1953 by Van Cleef & Arpels, which bought it back in 1960. In 1964, Harry Winston set it in a ring that he sold to Marjorie Merriweather Post, who bequeathed it to the Smithsonian Institution, Washington, D.C., where it is exhibited in company with an even more famous blue diamond, the Hope diamond.

Opposite: Corsage brooch in platinum and diamonds. Cartier Paris, 1912
This jewel was conceived for a pear-shaped diamond (E, VS1) of 34.25 carats—34.08 after it was recut by Cartier—two navette-cut diamonds, one of 23.54 and the other of 6.5 carats, and a heart-shaped diamond of 3.53 carats. Two rows of linked diamonds set as lilies of the valley are held at either end by two elements set with brilliant-cut diamonds. The two rows meet in a heart-shaped diamond, from which hangs the pear-shaped diamond and the two navette-cut diamonds, all surrounded by more diamonds set as lilies of the valley.

The jewel was made on special order for S. B. Joël, owner of the four major stones used in the piece, which came from his diamond mine in South Africa. This splendid example of the Belle Époque was sold by Christie's, Geneva, for 3,850,000 Swiss francs on 16 May 1991.

deciphering of the Rosetta Stone (1822), the opening of the Suez Canal (1869), the premiere of Verdi's opera Aïda (1871), and the discovery of Tutankhamen's tomb (1922). The theme of Egypt, of course, served as a source of inspiration for the decorative arts as well.

Right: *Design for a "modern"-style corsage brooch, ink and watercolor on tracing paper, 4³/₄ x 2" (12.1 x 5 cm). Archives Cartier Paris, 1913 The design specifies an 11.9-carat triangular emerald with flattened corners, four emerald drops, six cabochon emeralds, diamonds, sixteen pearls, and onyx cut in rounds, squares, and triangles.*

Opposite: *Studies of several Egyptian ornamental motifs, 9⁵/₈ x 6⁷/₈" (24.5 x 17.5 cm). Archives Cartier Paris, ca. 1910*

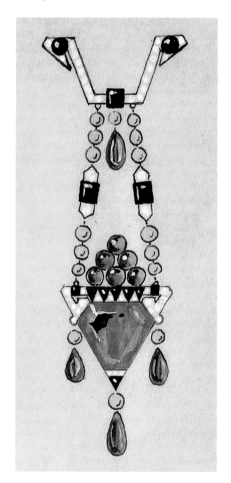

*U*nder the creative guidance of Louis Cartier and his designers, in 1907–8 the Garland style began to move toward a more restrained, more geometric design, flavored by touches of color.

Above: *Brooch in platinum, emeralds, diamonds, pearls, and onyx. Cartier Paris, 1913. Cartier Collection This brooch was executed from the drawing at right. Cartier's archives note that later that year, the emeralds of this brooch were removed and sent to Cartier New York, where the jewel was remounted following the original design. The setting left in Paris was used for an emerald of 4.49 carats, the space left by the larger triangular emerald being adeptly filled by diamonds and onyx motifs.*
Egyptian art had long fascinated Europe, especially France, sparked by a series of events such as Napoleon's campaign in Egypt (1798), the

Tribus Sauvages.

Lotus croissant dans le nil

Représentation égyptienne de la plante du papyrus, complet du type chapiteau, du fût et de le base des colonnes égyptiennes.

Eventail

Représentation égyptienne du lotus avec des boutons.

natte égyptienne.

Pourvernail orné du lotus et de l'œil, représentant la divinité.

Pourvernail d'autre genre.

*O*bviously, Cartier did not restrict its use of platinum to jewelry. It made many watches for women (pendant watches, watch brooches, and bracelet watches) and for men (pocket watches or wristwatches with leather straps) in platinum, plain or decorated with enamel, pearls, or diamonds. For such objects as cases in semiprecious stones, platinum constituted the ideal base and could ensure the strength of catches set with gems.

Above left: *"Santos-Dumont" watch in platinum. Cartier Paris, ca. 1910. Cartier Collection*

Above right: *Pendant watch in platinum and diamonds, front view (above) and back view (below). Cartier Paris, ca. 1910. Collection Cartier The double border, monogram, and watch ring are set with rose-cut diamonds. The watch hangs from a diamond and pearl sautoir.*

Opposite above left: *Regency-style watch brooch in platinum, brilliant-cut and rose-cut diamonds, and royal blue enamel. Cartier London, 1909. Cartier Collection*

Opposite above right: *Cigarette case in platinum, agate, sapphires, and diamonds. Cartier Paris, 1912. Cartier Collection*

The agate case has a platinum catch set with sapphires and rose-cut diamonds. The monogram is in platinum and rose-cut diamonds. The original owner was Lady Henriette Acton-Mitchell.

Below left: *Bracelet watch for women in platinum, rock crystal, and diamonds. Cartier Paris, 1913. Cartier Collection*
This round watch is framed by an oval plaque of engraved rock crystal set with rose-cut diamonds.

Below right: *Bracelet watch for women in platinum, diamonds, and silk. Cartier Paris, ca. 1913. Cartier Collection*
This square watch with sharp right angles has a bezel set with rose-cut diamonds and yokes with eight collet diamonds attached to the silk moiré strap by means of two bars set with rose-cut diamonds.

The most precious and noble jewel, as it was worn above all the others, was the head ornament, which often had a considerable weight in precious stones to support. Platinum's strength, which allowed it to be worked in reduced dimensions, resulted in tiaras that were less heavy. In addition, platinum's resistance to corrosion (another of its fundamental properties) makes it ideal as a setting for diamonds.

Below: *Kokoshnik-style tiara in platinum, diamonds, and pearls. Cartier Paris 1908. Cartier Collection*
Four rows of diamonds and two of pearls (in alternating sizes) meet at the ends of the tiara. In the center, fifteen graduated pear-shaped diamonds, with the largest in the center, are suspended from caps set with diamonds. Pearls alternate with the pear-shaped diamonds.

*T*iara in platinum, blue steel, diamonds, and rubies. Cartier Paris, 1914. Cartier Collection
Nine pear-shaped diamonds with a total weight of 20.05 carats framed by small calibré-cut rubies are set in a flexible steel band studded with thirty-six diamonds and rimmed with a band of diamonds and one of calibré-cut rubies. This tiara was made on special order for Madame Marghiloman, the sister of Queen Marie of Romania.

ART DECO
1915–1925

The style characterized by the curved line evolved into new geometric forms that influenced all the decorative arts of the period. Cartier, which had forecast Art Deco in its work, endowed it with a sense of balance and an unequaled originality. In the domain of jewelry, platinum catalyzed this creative evolution, always with the aim of conferring ever more splendor and beauty on precious gems.

The Galérie des Boutiques at the Exposition Internationale des Arts Décoratifs et Industriels Modernes, Paris, 1925. The first international exhibition of decorative arts, held in Paris in 1925, fed the current enthusiasm for the applied arts, extolled a refined art of living, and marked the height of the Art Deco style.

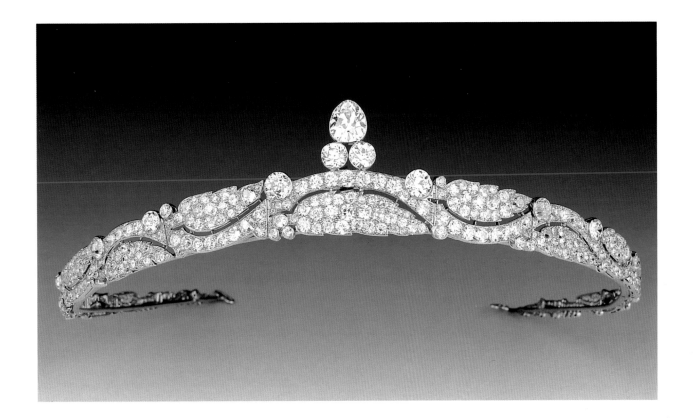

World War I barely affected Cartier's employment of platinum. Cartier's relationship with Saint Petersburg was interrupted in 1914, but Cartier New York had opened its doors in 1909 (the same year that the London branch moved to 175–76 New Bond Street), eight years before it established its present residence in the former Morton Plant mansion at the corner of Fifth Avenue and Fifty-second Street. Business did not suffer, and the evolving Art Deco style favored the use of platinum as the setting for the precious stones.

Opposite: *Snake necklace in platinum and diamonds. Cartier Paris, 1919. Cartier Collection*
This flexible necklace represents a snake with its

head and tail intertwined. A triangular diamond hangs from its tongue, composed of a baguette-cut diamond.

Above: *Tiara in platinum and diamonds. Cartier Paris, 1919. Cartier Collection*
The band displays a leaf motif in pavé diamonds crowned with three central diamonds. The tiara contains 544 diamonds with a total weight of 54.61 carats.

Below: *Bracelet in platinum and diamonds. Cartier Paris, 1923. Cartier Collection*
One of the two circular motifs functions as the clasp.

O pposite: *Pendant necklace in platinum, diamonds, emeralds, and engraved rock crystal. Cartier Paris, 1921*
A cabochon emerald of 15.04 carats is mounted in a plaque of rock crystal framed by diamonds. The pendant is attached to a black silk cord embellished with a stylized four-leaf clover and suspending two emeralds crowned with diamond caps.

Above: *Watch clip in platinum, gold, and diamonds. Cartier Paris, ca. 1920. Cartier Collection*
The clip is gold.

Above left: *Extra-flat pocket watch with platinum cover. Cartier Paris, 1919. Cartier Collection*

Below: *Extra-flat pocket watch with platinum cover. Cartier Paris, 1919. Cartier Collection*
The cover is engraved with the coat of arms of Alfred Potocki.

Left: *Square Regency-style watch in platinum, onyx, and diamonds on a silk moiré band. Cartier Paris, 1919. Cartier Collection*

Above left: *Square platinum wristwatch. Cartier Paris, ca. 1915. Cartier Collection*

Above right: *"Turtle" platinum wristwatch. Cartier Paris, 1919. Cartier Collection*

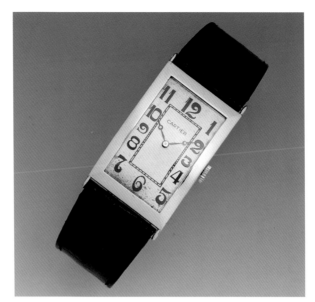

Above left: *Regency-style watch in platinum, onyx, rose-cut diamonds, and silk moiré. Cartier Paris, 1919. Cartier Collection*

Above right: *Rectangular wristwatch with curved case in platinum. Cartier Paris, 1921. Cartier Collection*

Left: *Bracelet watch in platinum, onyx, rose-cut diamonds, and gold. Cartier Paris, 1921. Cartier Collection*

*T*he vogue for Eastern art found expression mainly in objects and accessories: vanity cases, compacts, evening bags, ornamental boxes, and cigarette cases.

Above: *Design for a cigarette case in semiprecious stone, the central and two side arabesque motifs in turquoise enamel, black enamel, gold, platinum, and diamonds. Archives Cartier Paris, ca. 1920*

Opposite: *Cigarette case in platinum, nephrite, diamonds, sapphires, and black enamel. Cartier, ca. 1920. Cartier Collection*
The rectangular nephrite case is decorated with a miniature in an oval frame of rose-cut diamonds and calibré-cut sapphires accented with two arrows of rose-cut diamonds and two cabochon sapphires rimmed in black enamel. The catch and the two hinges are set with rose-cut diamonds and cabochon sapphires.

Left: *Brooch in platinum, engraved turquoise, diamonds, and black enamel. Cartier Paris, 1920. Cartier Collection*

Right: *Brooch in platinum, diamonds, and pearls. Cartier Paris, 1923. Cartier Collection*
The flow of the pearls is broken up by diamonds. The suspended pearl weighs 111.28 grains.

Below left: *Brooch in platinum, diamonds, and turquoise. Cartier Paris, 1920. Cartier Collection*

Below right: *Table bell in platinum, onyx, turquoise, and diamonds. Cartier, ca. 1920. Cartier Collection*
The cabochon turquoise, on a platinum base set with rose-cut diamonds, serves as the push button. It rests on an onyx sphere.

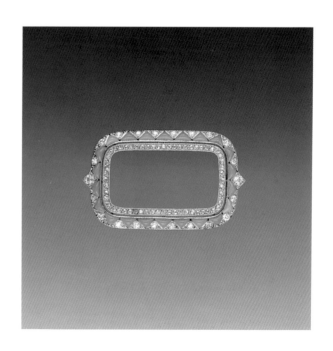

*I*n the 1920s, Cartier used Chinese jades dating from the second half of the nineteenth century, inserting them into jewels or objects or making them the base for table clocks, especially the famous mystery clocks.

Right: *Pendant in platinum, jade, rubies, and diamonds. Cartier Paris, 1921. Cartier Collection*
The piece of carved jade is set with a cabochon ruby, and the finger ring attached to the pendant is of cabochon rubies and diamonds.

Below: *Jabot pin in platinum, jade, diamonds, sapphires, black enamel, and gold. Cartier Paris, 1924. Cartier Collection*
The carved jade chimera has rose-cut diamond eyes rimmed in black enamel and is decorated with a diamond and cabochon sapphire motif. The support is in rose-cut diamond and black enamel.

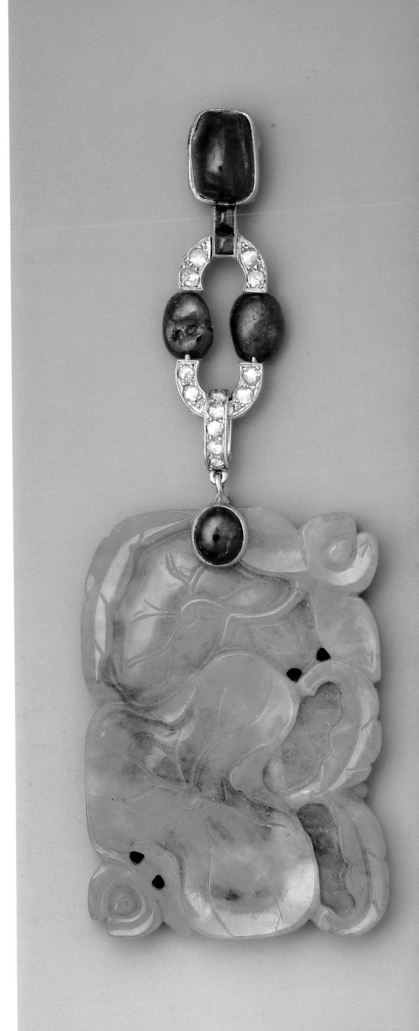

The kingdom of precious stones is a paradise of colors that have remained unchanged for centuries. Platinum gives the creative jeweler an ideally discreet base from which to put into play the subtle color relationships of precious stones. Louis Cartier found the combination of blue and green especially pleasing, and he used it often. It had already appeared combined in pieces of enamel at the beginning of the century, and he paired it by means of sapphires and emeralds, lapis lazuli and nephrite.

Above and opposite (detail): *Necklace in platinum,* *sapphires, emeralds, and diamonds. Cartier Paris, 1922*

Two strands of sapphires are fastened by a clasp set with a cabochon emerald. They are connected with strands of emerald beads by means of two diamond links attached with smaller links of calibré-cut sapphires. The two strands meet in an oval sapphire of 40.35 carats, which suspends a drop emerald of 65.57 carats by means of a motif composed of links of diamonds, calibré-cut sapphires, a double row of emerald beads, a cabochon sapphire, and a cap of calibré-cut sapphires.

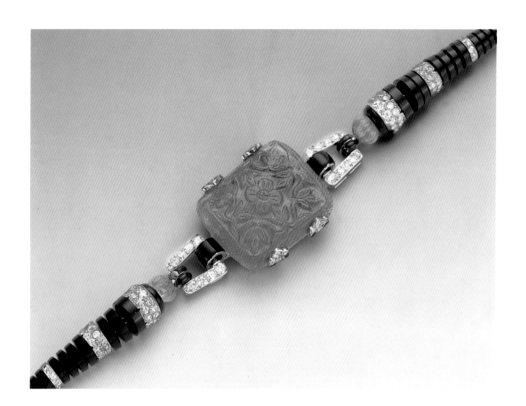

Above: *Bracelet in platinum, emeralds, diamonds, onyx, and black enamel. Cartier Paris, 1922. Cartier Collection*
The central emerald is engraved.

Right: *Bracelet in platinum, emeralds, onyx, and diamonds. Cartier Paris, 1923. Cartier Collection*
The bracelet is composed of emerald beads, onyx tubes, and diamond motifs set with cabochon onyx on platinum.

Opposite above: *Ring in platinum, diamonds, onyx, calibré-cut coral, and an emerald. Cartier Paris, 1922. Cartier Collection*
The cabochon emerald is framed in onyx within a rectangle set with a diamond at each corner. The sides are studded with diamonds and a calibré-cut coral.

This modern-style corsage brooch is attached at the top, where a pear-shaped cabochon emerald branches into a diamond limb shaded in onyx and ending in two round cabochon emeralds. From it hang two rings in diamonds and onyx, each closed by an accent in onyx set with a square-cut emerald, the lower one suspending two pendants with emerald beads capped with onyx and diamonds. At the bottom of these elements are two small arrows set with rose-cut diamonds.

Below: Bracelet in platinum, emeralds, coral, diamonds, and black enamel. Cartier Paris, 1922. Cartier Collection
Emerald and coral beads alternate with diamond rondelles rimmed with black enamel on platinum.

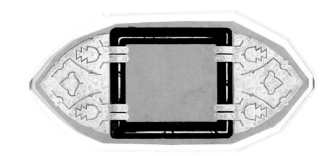

*O*ne of the strongest expressive elements of Art Deco is the opposition it sets up between black and white, a contrast that platinum accentuates.

The drawings found in Cartier's archives in Paris bear witness to the wealth of inspiration of the company's various designers, who only rarely initialed their ideas. Cartier's organization fostered the sense of a team working to create jewelry, which was not considered final until it received the notation "A Ex. L. C.," that is, "to execute, Louis Cartier."

Right: *Designs for brooches. Archives Cartier Paris, ca. 1925*

Opposite above: *Brooch in platinum, diamonds, and black enamel. Cartier New York, ca. 1925. Cartier Collection*
A rectangular link is half pavé diamonds, half black enamel. At one end is a pear-shaped diamond framed with black enamel; the motif at the other end is identical but with the colors reversed. The links that hold them to the central link are also reversed.

Opposite center: *Brooch in platinum, rock crystal, diamonds, emeralds, and onyx. Cartier Paris, 1923. Cartier Collection*
A ring of rock crystal, rimmed on the inside with diamonds alternating with onyx, holds two symmetrical motifs realized in diamonds with onyx foliage accented with four cabochon emeralds and rimmed with calibré-cut onyx.

Opposite below: *Dragon brooch in platinum, rock crystal, diamonds, onyx, and black enamel. Cartier Paris, 1925. Cartier Collection*
The barrel-shaped (in ellipse) rock crystal link is rimmed on the inside with black enamel and holds two diamond motifs at either end, each attached with a diamond link. One motif represents the head, the other the tail of a stylized dragon, with an onyx eye.

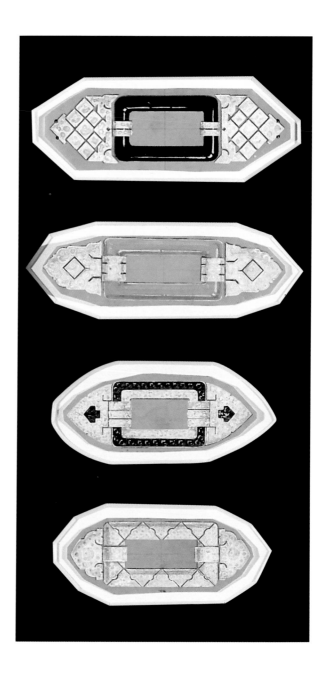

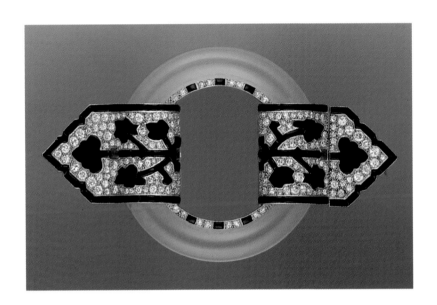

*A*lmost all of Cartier's Egyptian-style creations incorporated a piece of an original antique from the dynasties of the Pharaohs—a piece of blue faience, a portion of carved ivory or wood.
Above: Scarab brooch in platinum, blue faience, diamonds, onyx, emeralds, sapphires, ruby, and amethyst. Cartier London, 1924. Cartier Collection
The scarab is studded with cabochon gems: two emeralds, two sapphires, a ruby, and an amethyst. The zigzag border is diamond and onyx.

Above, below, and opposite top: *Winged scarab brooch and belt ornament in platinum, smoky quartz, emeralds, blue faience, diamonds, and black enamel. Cartier London, 1924. Cartier Collection Carved in smoky quartz, the scarab has two cabochon emeralds for eyes. Sections of the blue faience wings are set with pavé diamonds. The four cabochon emerald accents in the wings create a vivid contrast of colors. The border is black enamel.*

This brooch was the emblem of Cartier Collection at its first exhibition, "L'Art de Cartier," held at the Petit Palais, Paris, in 1989–90 at the request of the Musées de la Ville de Paris.

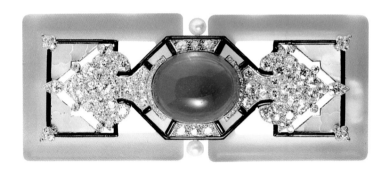

Against the cool and neutral background of platinum, pearls and mother-of-pearl, precious fruits of the sea, lost none of the nuances of their soft and iridescent sheen. They were often harmonized with precious stones carefully chosen for their colors.

Above: *Brooch in platinum, a cabochon sapphire of 57.63 carats, rock crystal, pearls, diamonds, and onyx. Cartier Paris, 1924. Cartier Collection*

Opposite above: *Necklace in platinum, diamonds, and pearls. Cartier New York, 1923*
The chain of variously cut diamonds could be separated into a necklace and bracelet. Two pearls, one white and one gray, hang from the chain.

Opposite below left: *Parakeet brooch in platinum, rubies, emeralds, diamonds, and a pearl. Cartier London, 1925. Cartier Collection*

Opposite below right: *Two rings in platinum, pearls, and baguette-cut diamonds. Cartier Paris, 1925. Cartier Collection*
It is thought that the two pearls came from a pair of earrings made for Queen Marie of Romania.

A *this time, onyx was already used mainly as a decorative element of trim or for its contrast with white or green, in the form of lines or small cabochon dots in a regular pattern. Starting in 1914, it gained a new application as irregular spots, which led to the famous panther theme.*

Above: *Cup brooch in platinum, onyx, diamonds, an oval cabochon emerald, and two pear-shaped cabochon rubies. Cartier Paris, 1925. Cartier Collection*
This brooch was originally owned by Mrs. William K. Vanderbilt.

Opposite above left: *Pendant watch in platinum, dial with Roman numerals, the back side (at right) decorated with the panther pattern of onyx spots on a ground of pavé diamonds. Cartier Paris, ca. 1920. Cartier Collection*

Opposite above right: *Safety pin brooch in platinum, onyx, and diamonds. Cartier Paris, 1920. Cartier Collection*

Opposite below: *Evening bag with clasp in platinum with the panther pattern in brilliant-cut and rose-cut diamonds and cabochon onyx. Cartier Paris, 1925. Cartier Collection*

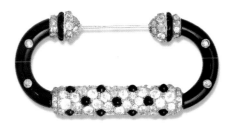

"*Sphinx*" *pendant necklace in platinum, onyx, rubies, emeralds, diamonds, black enamel, moonstone, and a pearl. Cartier Paris, 1921. Cartier Collection*

Within the onyx rectangle shaped as an Egyptian gateway is a double lotus flower motif in rubies and emeralds accented with a diamond in a triangle of onyx with emerald corners. Six collet diamonds form a fringe below the rectangle; above it is a band of eight triangular onyx alternating with diamonds. A diamond sphinx sits over the gateway, its head a carved moonstone, crowned with a motif in two emeralds and a ruby. A black enamel ring connects the pendant to the necklace through the intermediary of a motif featuring a pearl. The necklace contains a string of collet diamonds and onyx tubes rimmed with rose-cut diamonds.

Opposite above and below (design): *Egyptian head brooch in platinum, blue-green faience, diamonds, onyx, emeralds, rubies, and black enamel. Cartier New York, ca. 1925. Cartier Collection (design: Archives Cartier New York)*

The head, of blue-green Egyptian faience, wears a crown of diamonds, onyx, and a square-cut emerald. The side of the head is decorated with a motif of three rings set with diamonds and emeralds and rimmed in black enamel. The bust sits on a pedestal of stylized leaves in black enamel, pavé diamonds, emeralds, and calibré-cut rubies. The brooch was made on special order for Mrs. W. S. Moore.

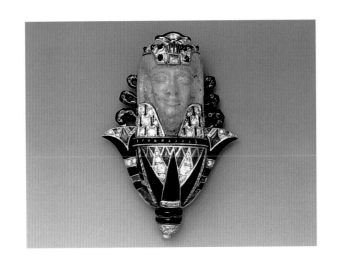

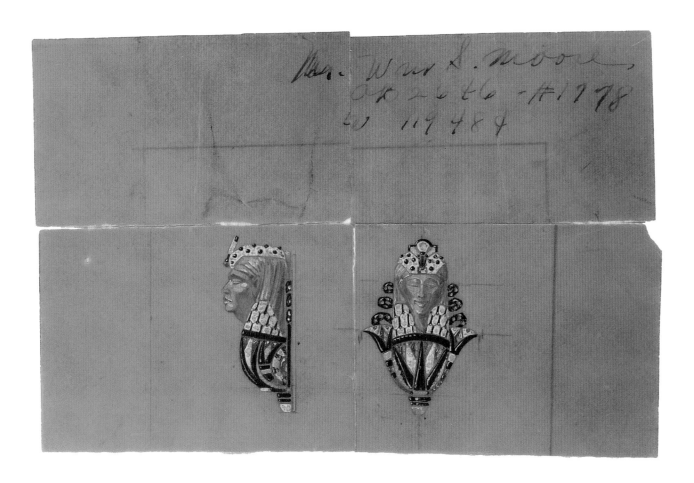

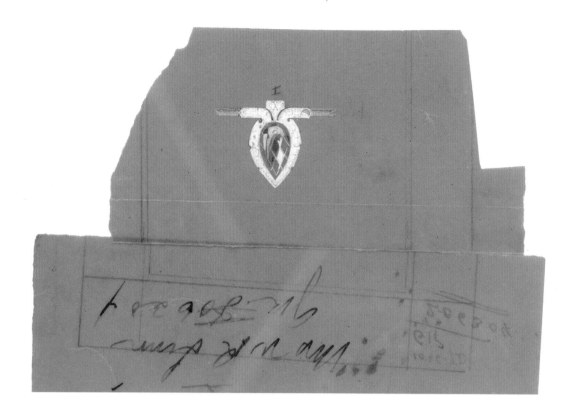

*C*artier's designers of fine jewelry were always prepared, as they remain today, to pay full attention to clients' personalized requests, making a jewel or a precious object from an idea or a yearning, turning it into a symbol of love or a commemoration of an event. The designs that carry clients' names such as those found on this page and the following offer clear examples.

Above and top (design): *Ring in platinum, pear-shaped briolette blue diamond, and pavé diamonds. Cartier New York, 1925 (design: Archives Cartier New York)*

Right and opposite (design): *Pendant earrings in platinum, emeralds, and diamonds. Cartier New York, 1927 (design: Archives Cartier New York)*

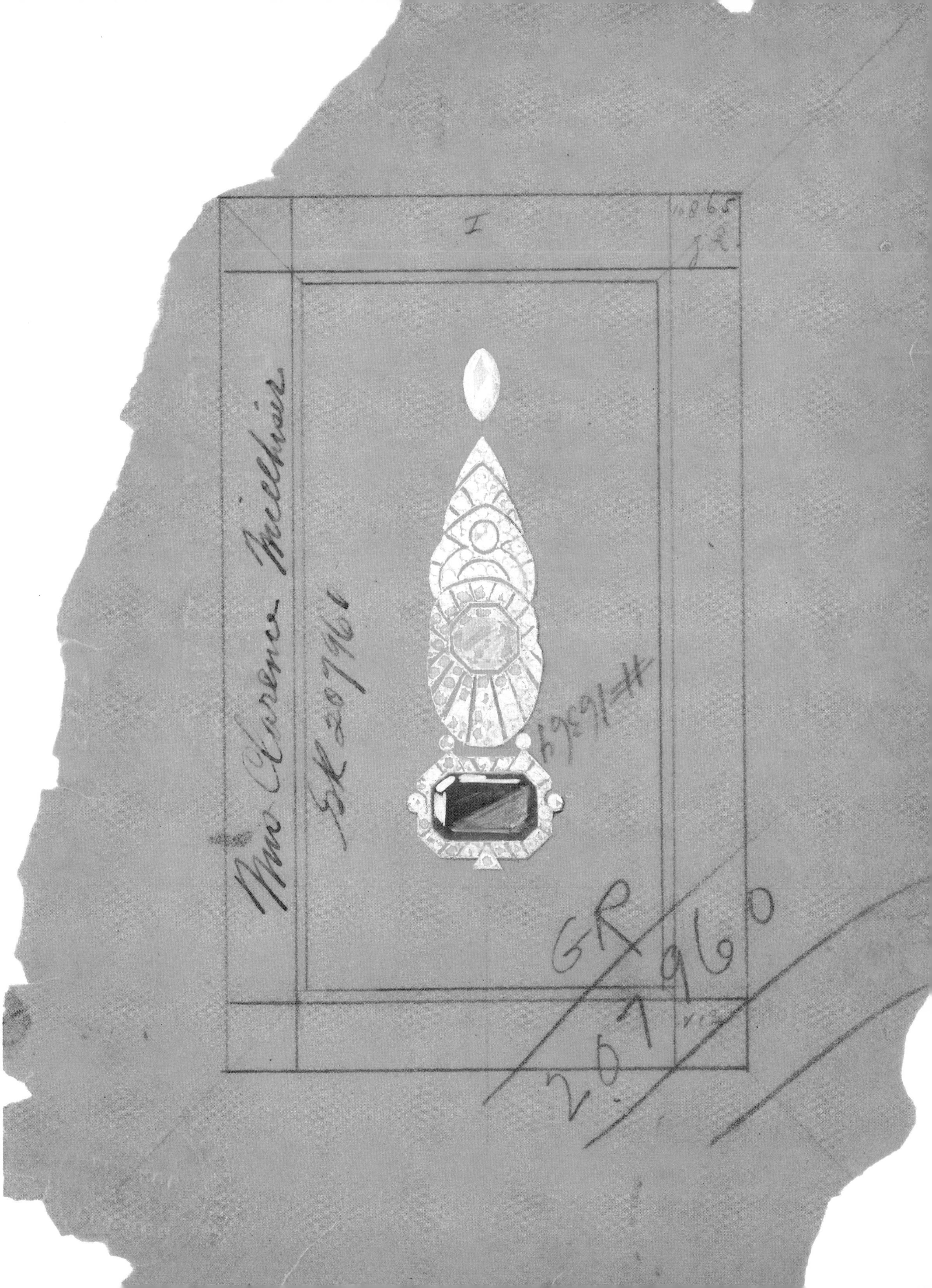

I

Mrs Clarence Mackhair

SK 207960

#16969

10865
J2

GR
207960
V13

In the 1920s, watches for women were almost always made in platinum, in white or accented with vivid colors.

Above: *Bracelet watch in platinum, diamonds, onyx, and pearls. Cartier Paris, 1924. Cartier Collection*
The oval case is set with diamonds and is attached to the bracelet, composed of nine rows of pearls, by means of onyx yokes.

Right: *Small bird pendant watch in platinum, diamonds, onyx, coral, and enamel. Cartier Paris, 1925. Cartier Collection*
The watch takes the form of a seal, the back and swivel hook enameled in the Persian style. It is suspended by a black silk cord that ends in a stirrup set with diamonds and calibré-cut onyx. The sliding ring from which the cord hangs, set in pavé diamonds and rimmed with onyx, is held by a pin decorated with two small birds in diamonds, coral, and onyx.

Opposite: *Bracelet watch in platinum, diamonds, onyx, and pearls. Cartier Paris, 1924. Cartier Collection*
The rectangular case is set with brilliant-cut and baguette-cut diamonds. The bracelet is made up of seven rows of alternating small and large pearls.

Right: *Design for a shoulder brooch with a long tassel. Pencil, ink, and watercolor on tracing paper. Archives Cartier Paris, 1922*
The brooch was made in platinum, diamonds, emeralds, pearls, and onyx.
Below: *Design for a brooch with two side tassels. Archives Cartier Paris, ca. 1925*
Opposite: *Tiara in platinum, diamonds, emeralds, and pearls. Cartier Paris, 1923. Cartier Collection. Nineteen engraved emerald drops with a total weight of 230.95 carats decrease in size on either side of the central emerald, which is set inversely from the other emeralds and crowned with two pearls. Between the emeralds are a larger pearl topped with a smaller pearl. A semicircle of diamonds ending in two emerald beads holds the emeralds and pearls.*

The tiara was bought by the distinguished conductor Sir Thomas Beecham and subsequently acquired by Sultan Sir Mohammed Shah, Aga Khan III, for his wife, the Begum Andrée.

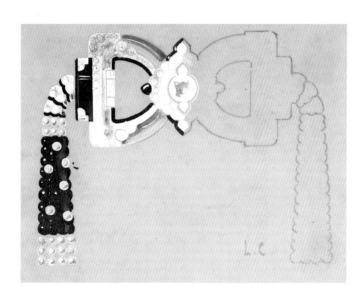

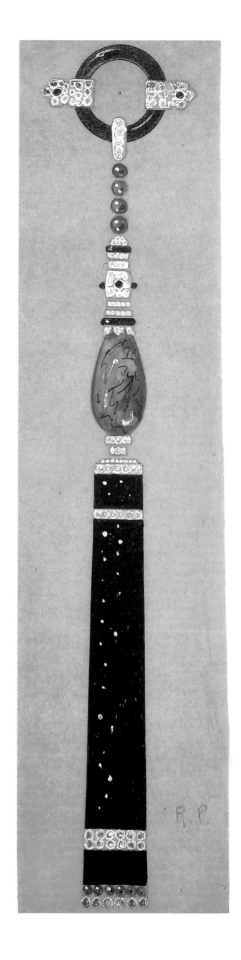

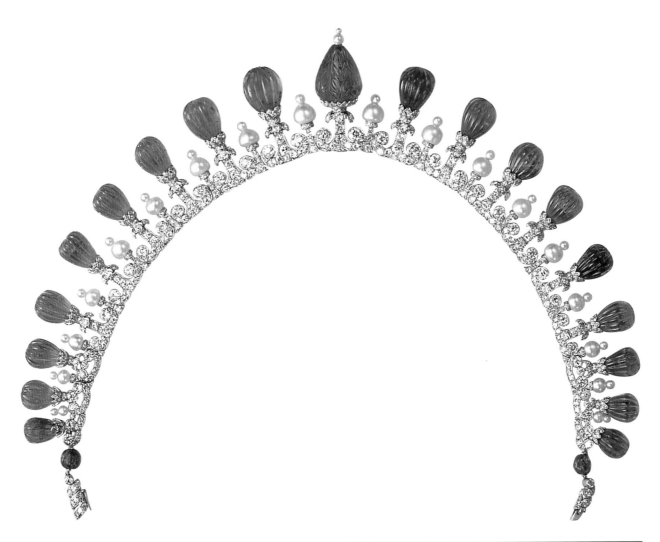

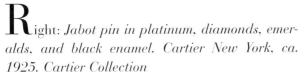 ight: *Jabot pin in platinum, diamonds, emeralds, and black enamel. Cartier New York, ca. 1925. Cartier Collection*
One of the two briolette emeralds is pear-shaped and rimmed in black enamel, the other is oval. The two black-enameled rings are free-moving. This pin was made on special order.

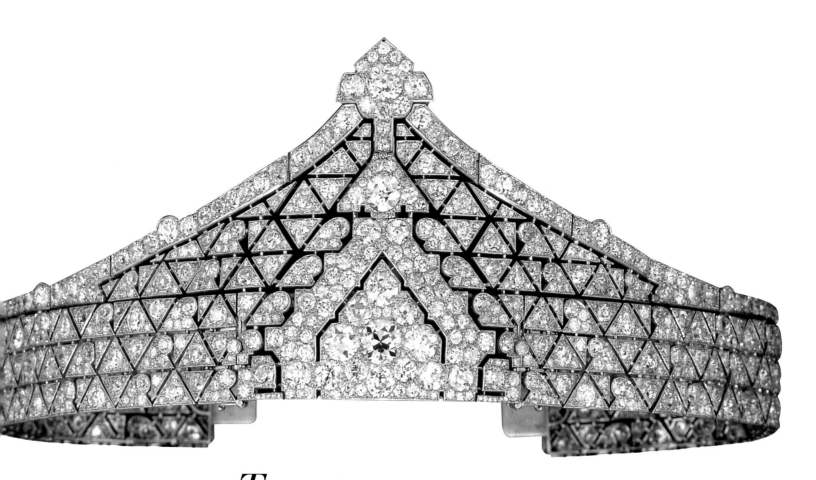

These two creations in platinum and diamonds were made at the request of clients who wished the jewels to incorporate as many of their diamonds as possible.

Opposite: *Bazu band bracelet, or armlet, in platinum and diamonds. Cartier Paris, 1922. Cartier Collection*
A flexible Persian-style plaque, which can be detached as a brooch, holds three leaf-shaped pendants. Made on special order, it uses 859 brilliant-cut diamonds with a total of 94.11 carats, and was restored with 28 brilliant-cut diamonds with a total of .99 carat.

Above: *Bandeau in platinum and diamonds. Cartier Paris, 1923. Cartier Collection*
The flexible mount is set with old mine-cut diamonds in an openwork, Eastern-style, geometric design. This head ornament, made on special order, can be separated into two bracelets and a brooch.

THE COLORS OF THE EAST
1926–1939

While the investigation of abstract forms led to the sublime perfection of white Art Deco, color and life exploded in the "tutti frutti" style and in the return to animal themes. The platinum of Cartier's jewels united East with West and form with color in a synthesis of pure art and lasting beauty.

The office of Marshal Louis-Hubert-Gonzalve Lyautey, designed by Eugène Printz, in the Musée des Arts Africains et Océanique, Vincennes.
The "Colonial Exposition" of 1931 at Vincennes—the origin of the museum—revealed to the world the French presence on five continents and asserted a colonial-inspired style.

*T*he stock market crash brough an end to the Roaring Twenties in both the United States and Europe. After that, fantasy survived only in the movies. The styles of this period stripped down to the bone; after the dazzling chromatic contrasts of colored gems, the final stage of Art Deco changes to all white.

Tortoiseshell began to be used for various accessories, and the solidity of platinum protected and enriched the fragile material.

Above: *Head ornament in platinum, tortoiseshell, diamonds, and pearls. Cartier Paris, 1923. Cartier Collection*
The ornament, which takes the form of a comb, carries a band of diamonds in a wavy openwork motif. It is edged on the top by another band of diamonds and bordered top and bottom by a row of pearls.

Opposite above: *Carrying comb in yellow tortoiseshell with case in platinum and diamonds. Cartier Paris, ca. 1928. Cartier Collection*
The comb is rimmed in platinum set with brilliant-cut and baguette-cut diamonds.

Right: *Design for a bracelet in platinum and diamonds. Archives Cartier New York, ca. 1925*
This was a special order from a client.

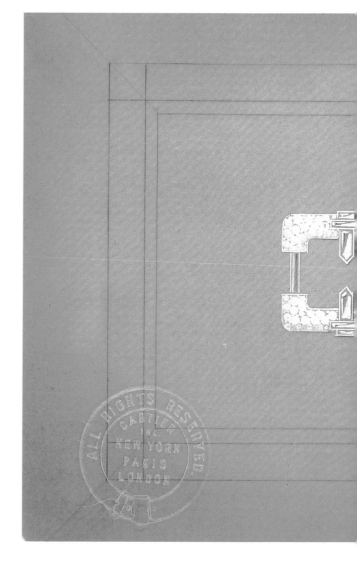

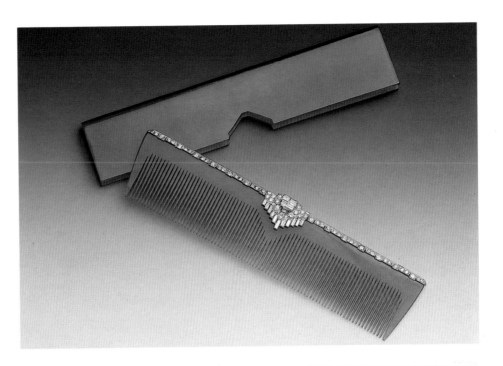

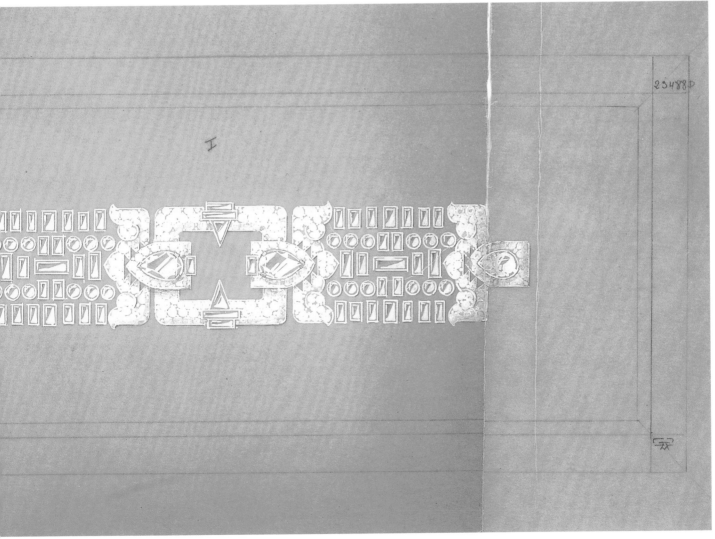

Above, below, and opposite: *Necklace in platinum, emeralds, and diamonds. Cartier London, 1926* The largest of the seventeen emeralds in this necklace weighs 70.24 carats. Brilliant-cut, baguette-cut, and half-moon-cut diamonds compose the chain. The necklace was bought by the maharaja of Nawanagar.

147

The simple and austere lines of diamonds set in platinum composed many small brooches with a figurative architectural theme (temple, pavilion, pagoda) or in the shape of a cup or urn. These jewels, all in white, prefigured the third and last phase of Art Deco.

Below: *Amphora and birds brooch in platinum, diamonds, and black enamel. Cartier Paris, ca. 1928. Cartier Collection*
The black enamel underlines the diamonds cut in various shapes.

Left and below: *Four brooches in platinum and diamonds. Archives Cartier Paris, 1927*
The diamonds are cut in various shapes, including brilliant, baguette, trapezium, square, and half-moon.

*S*educed by the qualities of the new metal, the maharajas forsook their old settings in gold in order to have their jewels remounted in platinum. Necklaces, turban ornaments, and other sumptuous jewels, set with fabulous stones (emeralds, rubies, sapphires, and diamonds from the Golconda mine), made their way to the workshops of Cartier London to be rejuvenated.

Opposite: *An invoice of 1928 sent from Cartier London to the maharaja of Nawanagar for the resetting of one turban ornament in platinum, using a pink diamond weighing 24.81 carats, and of another using an emerald of 80.88 carats. Archives Cartier London*

H.R.H. THE PRINCE OF WALES H.M KING OF SPAIN H.M. KING OF BELGIUM H.M.THE KING OF ENGLAND H.M.KING OF PORTUGAL H.M. KING OF ITALY H.M KING OF SIAM

IMPORTERS OF PEARLS

CARTIER LTD

Cartier S.A
13 Rue de la Paix Paris
Cartier Inc
653 Fifth Avenue New York

Please make Cheques
payable to Cartier Ltd.

J.C.   175-176 NEW BOND ST.
TELEPHONES: GERRARD 3754 TO 3758

LONDON

His Highness,
The Maharajah Jamsahib of Nawanagar,

ACCOUNT NO. 6.   The Palace, Jamnagar, Kathiawar. *Debit*   *Credit*

1928.									
Novbr.20	Remounting in platinum your Highness's Aigrette using your Aigrette, part of the diamond Sarpaich Your Highness's pink diamond weighing 24.81-cts. and 30 brilliant collets weighing 121.15-cts. invoiced July 20th, 1928 and supplying the following stones:-								
	50 brilliants 10.70-cts.	£390.							
	2 " 2.55-cts.	312.							
	2 baton " 0.37-cts.	17.							
	33 " " 4.75-cts.	230.							
	1 " " 0.11-cts.	5.							
	9 " " 2.29-cts.	205.							
	19 " " 5.20-cts.	364.							
	1 Pentagon brilliant 0.48-cts.	45.							
	1 Hexagonal " 0.71-cts.	104.							
	1 Trapeze " 0.38-cts.	60.							
	Mounting in platinum.	455.	2187	-	-				
Dec. 14	Mounting your Highness's own rectangular emerald weighing 80.88-cts. as an interchangeable centre to the diamond and emerald drop Turban Ornament and supplying 6 baton diamonds weighing 4.23-cts.		390	-	-				
1929.									
Oct. 24	Credit. BY Payment. (Part of £20,000)					2577	-	-	
			2577	-	-	2577	-	-	
			2577	-	-				
				-	-	-			

paid.

After the art of Islam, that of Egypt and China inspired Louis Cartier in turn. The clarity of diamonds on platinum was amplified by its contrast on a black ground. Vanity cases and cigarette cases became restrained ideal mediums for the artistic expression. The shapes of the cut stones (rose, brilliant, baguette, and so on) introduced contour to the otherwise flat planes of these accessories.

Above left: "Pagoda" vanity case in black enamel on gold, brilliant-cut, rose-cut, square-cut, and baguette-cut diamonds on platinum. Cartier Paris, 1927. Cartier Collection

Above right: "Déesse Maat" vanity case in black enamel on gold, rose-cut and baguette-cut diamonds on platinum. Cartier Paris, 1928. Cartier Collection

Precious colored "Indian" stones in fantasy cuts (engraved leaves, beads set with a diamond, ribbed beads, or simple cabochons) lent themselves admirably to the creation of brooches and bracelets with multicolored leaves, their veins sketched in brilliant-cut, square-cut, or baguette-cut diamonds. In the United States, these concoctions were joyously called "fruit salad" or "tutti frutti"

jewelry, names still used for Cartier's multicolored Art Deco creations.

Left: Ring in platinum, rubies, sapphires, diamonds, an emerald, and black enamel. Cartier Paris, 1927. Cartier Collection

An engraved cabochon emerald of 31.33 carats on a black enamel base is surrounded by two cabochon rubies, two sapphires, and diamonds.

Below: Brooch in platinum, rubies, sapphires, diamonds, and an emerald. Cartier Paris, 1930. Cartier Collection

The engraved emerald in the center sits between two sapphire bars and three ruby beads studded with a diamond. The brooch finishes at each end with a diamond motif. A series of hinges makes it convertible.

Opposite: Bracelet in platinum, emeralds, engraved sapphires, diamonds, and onyx. Cartier Paris, 1930

The central rectangular emerald of 76 carats is engraved with a sura, a chapter from the Koran. This bracelet, a special order from Sultan Sir Mohammed Shah, Aga Khan III, offers a sublime example of contrasting green and blue to which Cartier was partial.

The economic crisis set off in 1929 began to move in the direction of the Old World. In its wake, the Art Deco style evolved into its purest form: white on white. To the cold glitter of diamonds cut in geometric shapes (baguette, emerald, trapezium, or triangular cuts), rock crystal, polished and smooth as ice, added its transparency, both mounted in silver-white platinum.

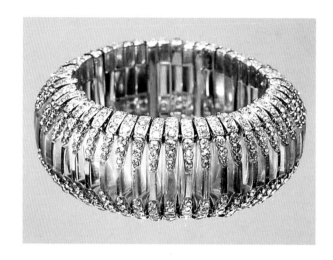

Right: *Pair of bracelets in platinum, rock crystal, and diamonds. Cartier Paris, 1930. Cartier Collection*
The diamonds are cut in brilliant and baguette shapes. The two bracelets, which have no clasp, can be slipped on by means of a clever mounting on platinum springs. The original owner, the actress Gloria Swanson, wore them together or separately, like two cuffs.

Opposite: *Choker in platinum and diamonds. Archives Cartier Paris, 1929*

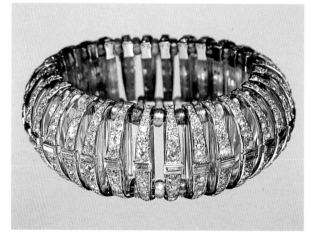

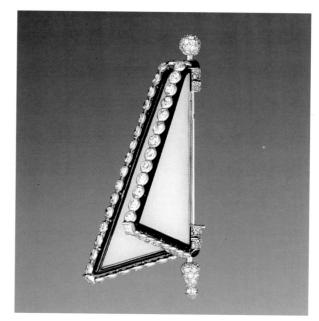

White Art Deco according to Cartier: platinum, diamonds, rock crystal, and white jade.
Above left: *Brooch in platinum, rock crystal, and diamonds. Cartier Paris, 1928. Cartier Collection Two palmette motifs in diamonds decorate an oval ribbed ring of rock crystal with two diamond links on either side.*
Above right: *Sail brooch in platinum, white jade, diamonds, and black enamel. Cartier Paris, 1930. Cartier Collection*

Two triangular sails in white jade are rimmed with black enamel and edged with collet diamonds. At the top and the bottom of the brooch are diamond spheres and a rose-cut diamond motif.
Below: *Bracelet in platinum, diamonds, and rock crystal. Cartier Paris, 1930. Cartier Collection Eight brillant-cut and baguette-cut diamond squares on a base of rock crystal are linked together by means of elements composed of a double row of baguette-cut diamonds.*

R ight: *Bracelet in platinum, rock crystal, and diamonds. Cartier Paris, 1934. Cartier Collection*
The engraved rock crystal is crowned by a large clasp in platinum set with brilliant-cut and baguette-cut diamonds. The clasp can be detached and used as a brooch.

Below left: *Door knocker brooch in platinum, diamonds, rock crystal, and onyx. Cartier Paris, 1931. Cartier Collection*
The brooch is formed by a half disk of faceted rock crystal enclosed in a half tube of onyx. Six baguette-cut diamonds decorate the upper part; a cross motif on the sides; and a diamond pavé tongue in the center.

Below right: *Bow tie brooch in platinum, rock crystal, and diamonds. Cartier Paris, 1935. Cartier Collection*
The bow tie in carved rock crystal is knotted with pavé diamonds and decorated at both ends with a line of diamonds.

Opposite above: *Pair of "tutti frutti" clips in platinum, rubies, emeralds, diamonds, and black enamel. Cartier New York, 1929. Cartier Collection Cabochon rubies, engraved ruby leaves, and smooth and ribbed emerald beads studded with a diamond encircle a ruby bead studded with a diamond on a ground of pavé diamonds accented with black enamel lines.*

Opposite below: *"Tutti frutti" bracelet in platinum, rubies, sapphires, emeralds, and diamonds. Cartier New York, ca. 1930. Cartier Collection A pavé diamond band runs the length of this brace-let, which is thickly planted with engraved ruby leaves, sapphire beads, and smooth and ribbed emerald beads studded with a diamond.*

Below: *"Tutti frutti" evening bag in platinum, deerskin, diamonds, emeralds, rubies, and black enamel. Cartier Paris, 1929. Cartier Collection The handbag is in blue deerskin, the clasp in pavé diamonds, baguette-cut and square-cut diamonds, and floral motifs in emeralds and engraved ruby leaves. The catch is an engraved oval cabochon ruby framed in black enamel and diamonds. The handle links have diamond accents.*

"Daisy Fellowes" necklace in platinum, emeralds, rubies, sapphires, and diamonds. Cartier Paris, 1926–36. Cartier Collection
This unique necklace represents the apotheosis of the "tutti frutti" style. It contains engraved emerald, ruby, and sapphire leaves, smooth and engraved cabochon and bead sapphires, and brilliant-cut and baguette-cut diamonds. Eleven briolette sapphires dangle below the foliage, and two engraved cabochon sapphires form the clasp.

Starting with a row of sapphire beads, this necklace went through ten transformations in ten years, from the first drawing in 1926 to the final version in 1936. Reportedly, its owner, Daisy Fellowes, wore it just once, to a grand ball in Venice hosted by the millionaire Carlos de Beistegui.

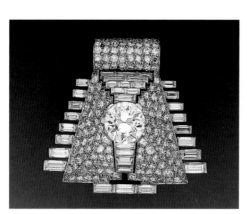

Top: *Pyramid clip in platinum and brilliant-cut and baguette-cut diamonds. Cartier Paris, 1935. Cartier Collection*

Above: *Door knocker double clip in platinum and brilliant-cut and baguette-cut diamonds. Cartier London, 1936. Cartier Collection*

Above right: *Egyptian necklace in platinum, blue faience, diamonds, onyx, and pearls. Cartier London, 1931. Cartier Collection*
The chain contains a double row of blue faience from Egypt and pearls. A triangular fragment of blue faience hangs from a diamond and onyx link, and the clasp is also of diamonds and onyx.

Below right: *Spiral brooch in platinum and brilliant-cut, square-cut, and baguette-cut diamonds. Cartier London, 1935. Cartier Collection*

Above: *Watch clip in platinum, an engraved emerald, and diamonds. Cartier Paris, 1938. Cartier Collection*

Top left: *"Tutti frutti" bracelet watch in platinum, diamonds, and rubies. Cartier Paris, 1926. Cartier Collection*

Top center: *"Tutti frutti" bracelet watch in platinum, diamonds, emeralds, rubies, and sapphires. Cartier Paris, ca. 1930. Cartier Collection*

Top right: *Bracelet watch in platinum and diamonds. Cartier Paris, ca. 1927. Cartier Collection*

Right: *Watch clip in platinum and brilliant-cut and baguette-cut diamonds. Cartier London, 1931. Cartier Collection*

The purity of design and the restraint of platinum combine to emphasize the elegance of pocket watches and bracelet watches.

Above: *Round pocket watch in platinum and onyx with rose-cut diamond Roman numerals. Cartier New York, 1926. Cartier Collection*

Below: *Round pocket watch with three dials for different time zones, with a single wind stem to set all three times simultaneously. Cartier Paris, ca. 1928. Cartier Collection*

Opposite above left: *Square pocket watch with shaped corners in platinum, rock crystal, and rose-cut diamonds. Cartier New York, 1929. Cartier Collection*

Opposite above right: *Square mystery pocket watch with cut corners and off-center transparent dial. Cartier Paris, ca. 1931. Cartier Collection*

Opposite below left: *Wristwatch with a seconds counter. Cartier, 1937 (made for Cartier by Patek Philippe). Cartier Collection*

Opposite below right: *Tank bracelet watch with platinum bracelet of seven rows of overlapping links and deployant buckle. Cartier Paris, 1934. Cartier Collection*
The prince of Nepal originally owned this watch.

The mystery clock "Model A" was created in 1913. The perfect proportions of the rock crystal body resulted in an object so classic that it has been used through the years up to the present with numerous variations on every element: the choice of a semiprecious stone base, which hides the movement; and the dial, hands, and decoration of the vertical shafts that house the small axles. The axles turn the rock crystal disks to which the hour and minute hands are separately attached.

Opposite: *"Model A" mystery clock in platinum and rock crystal on an onyx base; diamonds separate the Roman numeral hours. Cartier New York, 1937. Private collection*

Below: *Cigarette case in platinum, emeralds, and diamonds. Cartier Paris. Private collection*
The pavé emerald sides contain sixty emeralds with a total weight of 20.08 carats accented with two baguette-cut diamond bands. The push button is made of a baguette diamond of .62 carat. Remarkably elegant, this object offers a unique example of smooth platinum on a large surface. The perfect polishing in itself represents a tour de force.

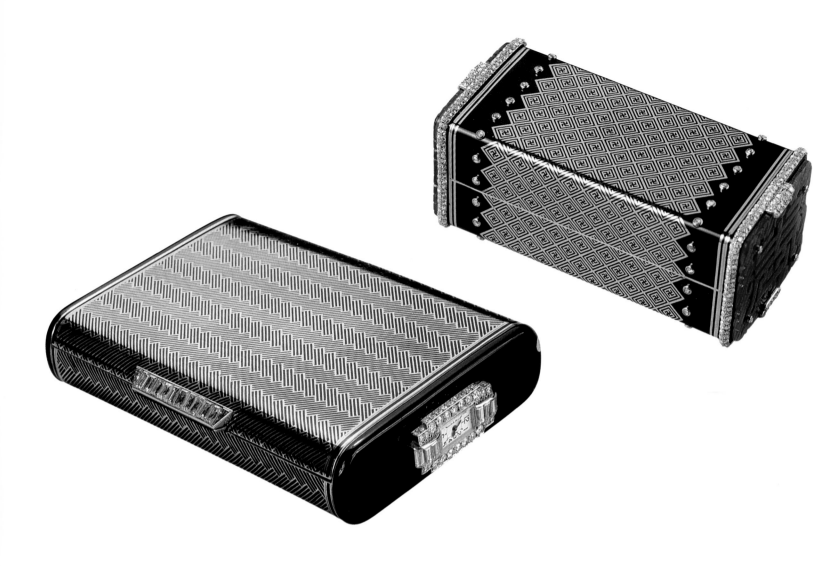

*P*latinum was often called on to serve as the base of an applied decoration such as an ornamental motif or a monogram, on gold vanity cases. Subsequently, the idea of an ornamental motif in platinum set in diamonds was transferred to gold jewelry.

Above left: *Vanity case in platinum, gold, black enamel, diamonds, and onyx. Cartier Paris, 1926. Cartier Collection*
The black enamel and gold case has a basket-weave pattern. The ends are in onyx, one of which holds a watch set with brilliant-cut and baguette-cut diamonds on platinum. The push button is set with diamonds on platinum.
Above right: *Vanity case in platinum, gold, black*

enamel, diamonds, and two Chinese plaques. *Cartier Paris, 1929. Cartier Collection*
The case is decorated in black enamel on gold in the swastika pattern, with two openwork Chinese plaques and pavé diamond and rose-cut diamond accents set in platinum.

Opposite: *Vanity case in platinum, gold, diamonds, and black enamel. Cartier London, 1935. Cartier Collection*
This large vanity case in gold is decorated with two diamond motifs on platinum rimmed in black enamel at either end and the monogram BG with the crown of a count on platinum in the center. Comtesse Granard (Béatrice) was the original owner.

Chimera bracelet in platinum, diamonds, sapphires, emeralds, and rock crystal. Cartier Paris, 1929. Cartier Collection
Two chimera heads, set entirely in diamonds, face each other, their eyes in pointed cabochon sapphires. Their noses and ears are in calibré-cut sapphires, jaws and wings in cabochon emeralds, and lower parts in ribbed rock crystal with two bands of square cabochon sapphires. The round and flexible bracelet is set with collet diamonds. This is the first chimera bracelet to be found in fine jewelry. Mrs. Blair Fairchild (Mrs. Havemeyer) was the first to own it.

*N*ext to the poetic and mysterious realm of fable, the world of sport also found expression in Cartier creations.

Left: *Skier brooch in platinum, brilliant-cut and baguette-cut diamonds, and calibré-cut rubies. Cartier London, 1930. Cartier Collection*

Below: *Design for a horse and jockey brooch, made on special order. Archives Cartier New York, 1940*

*C*artier also excelled in luxury accessories, often made with stones provided by clients, who wanted to update their old jewels or objects to the style of the day, including a setting in platinum.

Above left: *Belt buckle in platinum, onyx, and diamonds. Cartier Paris, 1930. Cartier Collection*
Two diamond bands placed in reverse tie together two rectangles half in onyx, half set with brilliant-cut and baguette-cut diamonds.

Above right: *Belt buckle in platinum, engraved rock crystal, diamonds, black enamel, and gold. Cartier Paris, 1927. Cartier Collection*
The belt is rimmed in black enamel, and the bars are of gold.

Right: *Belt buckle in platinum, diamonds, and onyx. Cartier Paris, 1928. Cartier Collection*
The half-moon buckle is decorated with stylized palmettes in diamonds and onyx accents. Its original owner was the maharani of Indore.

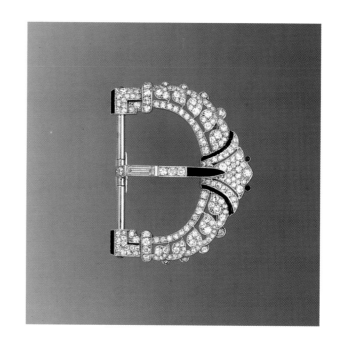

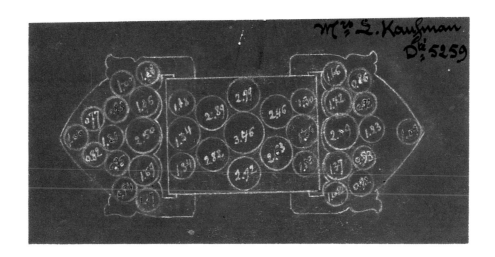

*D*esign for the project of transforming a rivière necklace into a brooch. Archives Cartier New York, 1928
This special order provides an exemplary demonstration of the designers' precision: the exact position of every stone has been marked to make sure that the new jewel in platinum has the right balance.

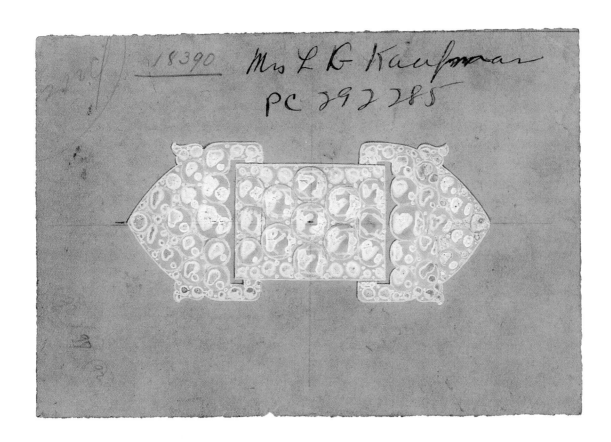

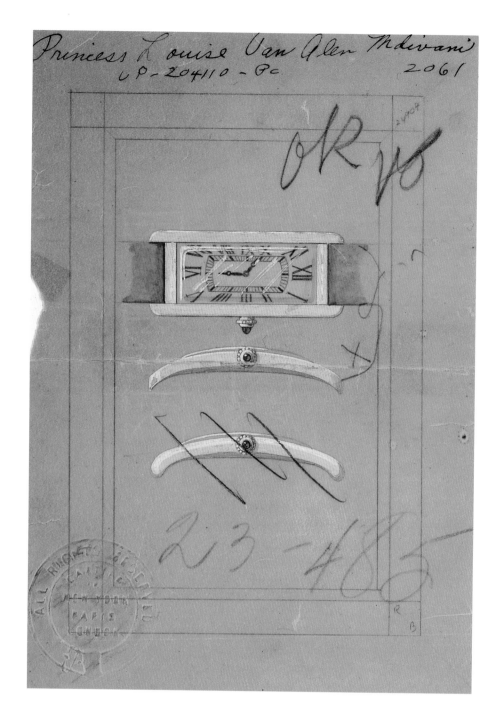

Above: *Design for a curved Tank watch in platinum. Archives Cartier, New York, 1935*
This was made on the special order of Princess Louise Van Alen Mdivani, sister-in-law of Barbara Hutton.

Opposite: *Octagonal mystery clock in platinum, rock crystal, black enamel, gold, coral, onyx, and diamonds. Cartier Paris, ca. 1926. Cartier Collection*
The dial is faceted rock crystal, the bezel in black enamel on gold edged in ribbed coral, the base onyx on gold. The various decorative elements— the Roman numerals, the hands shaped like a Chinese dragon, the letter F (for Farouk) and the flanking winged motifs—are all in rose-cut and brilliant-cut diamonds on platinum.

*P*latinum jewels belonging to the Duchess of Windsor: these were among the jewels sold by Sotheby's, Geneva, on 2 April 1987 to benefit the Institut Pasteur. This was surely the most spectacular sale of the century of a collection from a single and original source, many of which were made on special order in platinum by Cartier. The duchess, on principle, only wore platinum after five in the evening.

Above: *Bangle bracelet in platinum, rubies, and diamonds. Cartier Paris, 1938*
Two rubies from Burma with a total of 36.15 carats are framed in pavé and baguette-cut diamonds. The phrase "For our first anniversary of June third" is engraved inside.

Right: *Flamingo brooch in platinum, sapphires, rubies, emeralds, diamonds, and citrine. Cartier Paris, 1940. Private collection*
Pavé sapphires and square-cut rubies and emeralds form the bird's plumage. Its beak is set with a citrine and a cabochon sapphire, its eye with a round cabochon sapphire.

Opposite: *Necklace and bracelet in platinum, peridots, and brilliant-cut and baguette-cut diamonds. Cartier London, 1936*
Made on order for a client, they were sold by Christie's, Geneva, on 21 May 1992.

Platinum settings particularly enhanced colored gems, which had to be of exceptional quality. Platinum worked admirably for the design of graduated gems in a line. The splendor of the whole arose from the homogeneity of the colors of the assembled stones.

Below: *Bracelet in platinum, sapphires, and diamonds. Cartier London, 1938*
The bracelet incorporates five sapphires that originally weighed 59.49 carats and 187 baguette-cut diamonds weighing 52.38 carats. It was presented by Prince Johannes von Thurn und Taxis to his wife, Princess Gloria, née countess of Schönburg-Glauchau. On 17 November 1992, she sold it at auction at Sotheby's in Geneva for 539,000 Swiss francs.

Opposite: *Necklace in platinum, aquamarines, sapphires, and diamonds. Cartier Paris, 1939*
Five detachable brooches in aquamarines, sapphires, and brilliant-cut and baguette-cut diamonds are attached to a choker set with diamonds and square aquamarines.

Above: *Flower brooch in platinum, emeralds, rubies, and diamonds. Cartier Paris, 1940*
This brooch repeats the concept of the "tutti frutti" jewel, whose stones it reuses—nine engraved emerald leaves, ruby beads studded with a diamond, and collet diamonds. It forecasts the end of Art Deco and offers the new naturalistic theme, characterized by freeness of design and three-dimensional volumes.

Opposite: *"Merle Oberon" necklace in platinum, emeralds, and diamonds. Cartier London, 1938. Private collection*
This necklace, ordered by the actress Merle Oberon, contains twenty-nine baroque, polished emerald drops in graduated size, each studded with a diamond. A diamond cap connects them to the flexible platinum chain, composed of diamond rondelles. Oberon, of Irish, French, and Dutch extraction, was famous not only for her exotic beauty but also for her memorable interpretations in such films as Folies-Bergères *(1935) with* Maurice Chevalier *and* Wuthering Heights *(1939) with Laurence Olivier.*

In the 1930s, the tiara did not, as in the past, make up a significant part of Cartier's production. However, in those that were still made in London, for the English royalty and aristocrats, the stylistic traits of the period could be discerned, as well as the traces of faraway influences, such as Egyptian and Eastern art.

Above: *Tiara in platinum and diamonds. Cartier London, 1934. Cartier Collection*
The tiara is formed by a line of lotus flowers above a detachable bandeau in a geometric design. It was ordered by Sultan Sir Mohammed Shah, Aga Khan III, for his wife, the Begum Andrée.

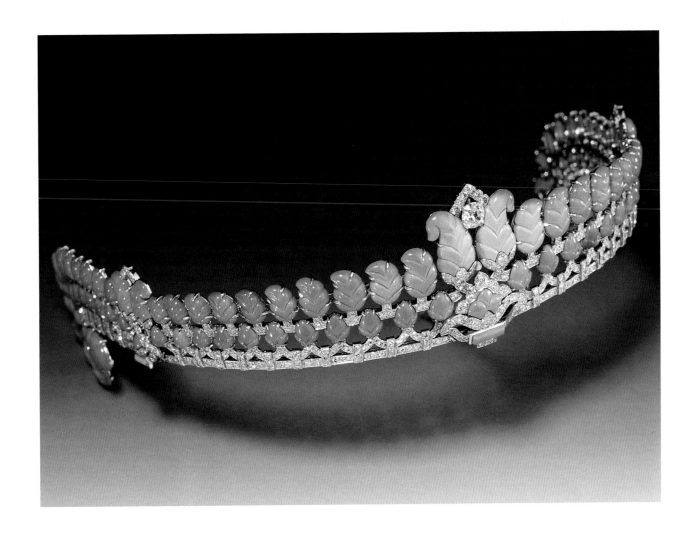

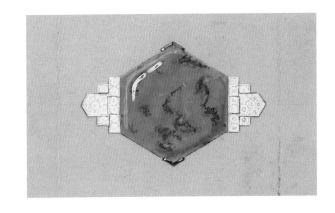

Above: *Tiara in platinum, turquoise ceramic, and diamonds. Cartier London, 1936. Cartier Collection*
Two rows of turquoise ceramic leaves of Kashmiri influence have a border of diamond motifs and are accented with a central and two side diamond motifs.

Right: *Design for a brooch in platinum, diamonds, and a hexagonal turquoise. Archives Cartier New York, 1930s*

FLORA AND FAUNA
1940–1967

Among the cultural disarray that followed the war, Cartier rediscovered his most original approach in a fresh and brilliant naturalism. Platinum, partly replaced by gold in the jewels of fantasy, appeared as the king of the night.

Neighborhood of the church of Saint Joseph in Le Havre (architect: Auguste Perret, ca. 1950). The urgency to reconstruct certain cities destroyed during World War II and the need to provide shelter for a growing population offered architects like Auguste Perret the opportunity to use reinforced concrete on a large scale.

*D*uring World War II, as platinum vanished from the jewelry workshops, its replacement, gold in all its colors of yellow, pink, and even white, became more familiar. Gold accommodated the use of fewer precious stones, which themselves were often replaced by what are called the semiprecious stones, notably citrine, a yellow stone in the quartz family. In the new creations in the naturalistic style, devoted to the flora and fauna, platinum became the support for ornamental diamond motifs, used to enhance the pistil of a flower or an animal's head.

Right: *Flower brooch in platinum, citrine, and diamonds. Cartier London, ca. 1940*

Below left: *Design for a flower with five petals in gold with pistil made of eight diamonds set in platinum. Archives Cartier Paris, ca. 1940*

Below: *Double flower brooch in platinum, gold, rubies, and diamonds. Cartier Paris, 1942*
Above four engraved gold leaves on a gold stem bloom two flowers made of gold and ruby beads, with platinum pistils set with diamonds.

Left: *Leaf brooch in platinum, gold, and diamonds. Cartier Paris, ca. 1945*
The gold leaf has smooth and braided threads. The central motif is set with a brilliant-cut diamond, baguette-cut diamonds in a star shape, and a garland of diamonds on platinum.

Center: *Leaf brooch in platinum, gold, and diamonds. Cartier Paris, ca. 1947*
The smooth veins of the gold leaf flow from the central diamond and are bordered by braided gold. The leaf is studded with small platinum and gold flowers.

Below left: *Flower bouquet brooch in platinum, gold, citrines, and diamonds. Cartier London, 1945*
The gold brooch offers light and dark citrine flowers, diamonds, and a diamond pavé platinum leaf.

Below right: *Flower bouquet brooch in platinum, gold, citrines, and diamonds. Cartier London, 1945*
This gold bouquet contains three light and dark citrine flowers and diamonds, tied with a diamond-set platinum element. Parts of the stems are set with diamonds on platinum.

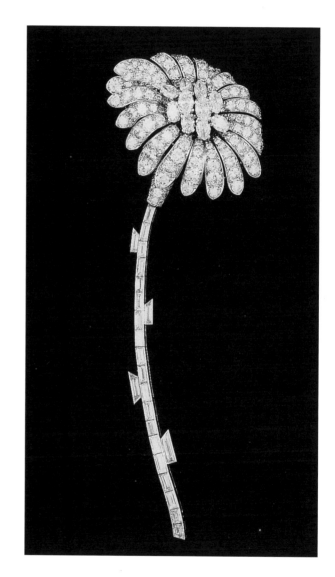

*W*hile gold decorated with platinum motifs formed a major part of postwar production, platinum jewels entirely set with precious stones did not disappear altogether. These, in an inversion, were enriched with refined elements set in gold.

Above: *Four-leaf clover brooch in platinum and diamonds. Cartier Paris, 1959*
Four pear-shaped diamonds are held on a stem of two baguette-cut diamonds.

Right: *Flower brooch in platinum and diamonds. Cartier New York, 1940*
Above a long stem set with baguette-cut diamonds nods a flower with distinct diamond petals and marquise-cut diamonds in its center. It was made on order for Lady Mendl (Elsie de Wolfe).

Opposite: *Flower brooch in platinum, gold, sapphires, diamonds, and emeralds. Cartier New York, 1941*
A long gold stem set with square-cut emeralds and two sapphire and diamond buds ends in a diamond pavé calyx bearing a large flower with wavy petals set with faceted sapphires and diamond motifs. The pistil is set with brilliant-cut and baguette-cut diamonds.

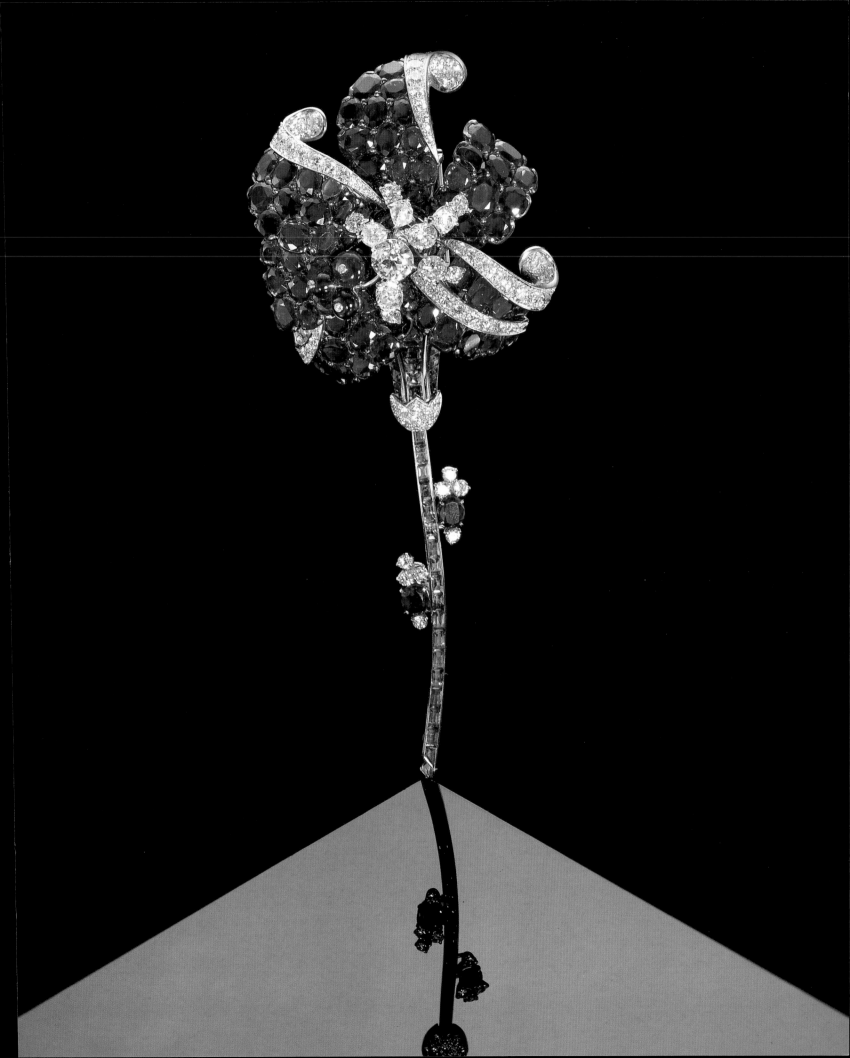

Right: Leaf brooch in platinum, emeralds, and diamonds. Cartier Paris, 1941
The Indian-style brooch contains ten emerald beads, each studded with a diamond, surrounded with brilliant-cut and square-cut diamonds and suspended from a diamond-pavé and baguette-cut diamond element.

Center: Pair of flower bouquet brooches in platinum, emeralds, sapphires, and diamonds. Cartier London, 1946
The two bouquets are set respectively with an emerald surrounded by square-cut sapphires and an emerald-cut sapphire surrounded by square-cut emeralds. The stems, in brilliant-cut and baguette-cut diamonds, are set with square-cut sapphires and two briolette emerald drops. They were made on special order.

Below left: Rose brooch in platinum, ivory, lapis lazuli, a square-cut sapphire, and diamonds. Cartier Paris, 1959

Below right: Flower brooch in platinum, diamonds, and an emerald. Cartier Paris, 1946
The pistil, composed of a pear-shaped cabochon emerald surrounded by brilliant-cut, baguette-cut, and square-cut diamonds, is held on a baguette-cut and brilliant-cut diamond stem. The brooch originally contained only diamonds.

Opposite: Rose brooch in platinum, coral, and diamonds. Cartier Paris, 1955
This brooch was bought by Daisy Fellowes.

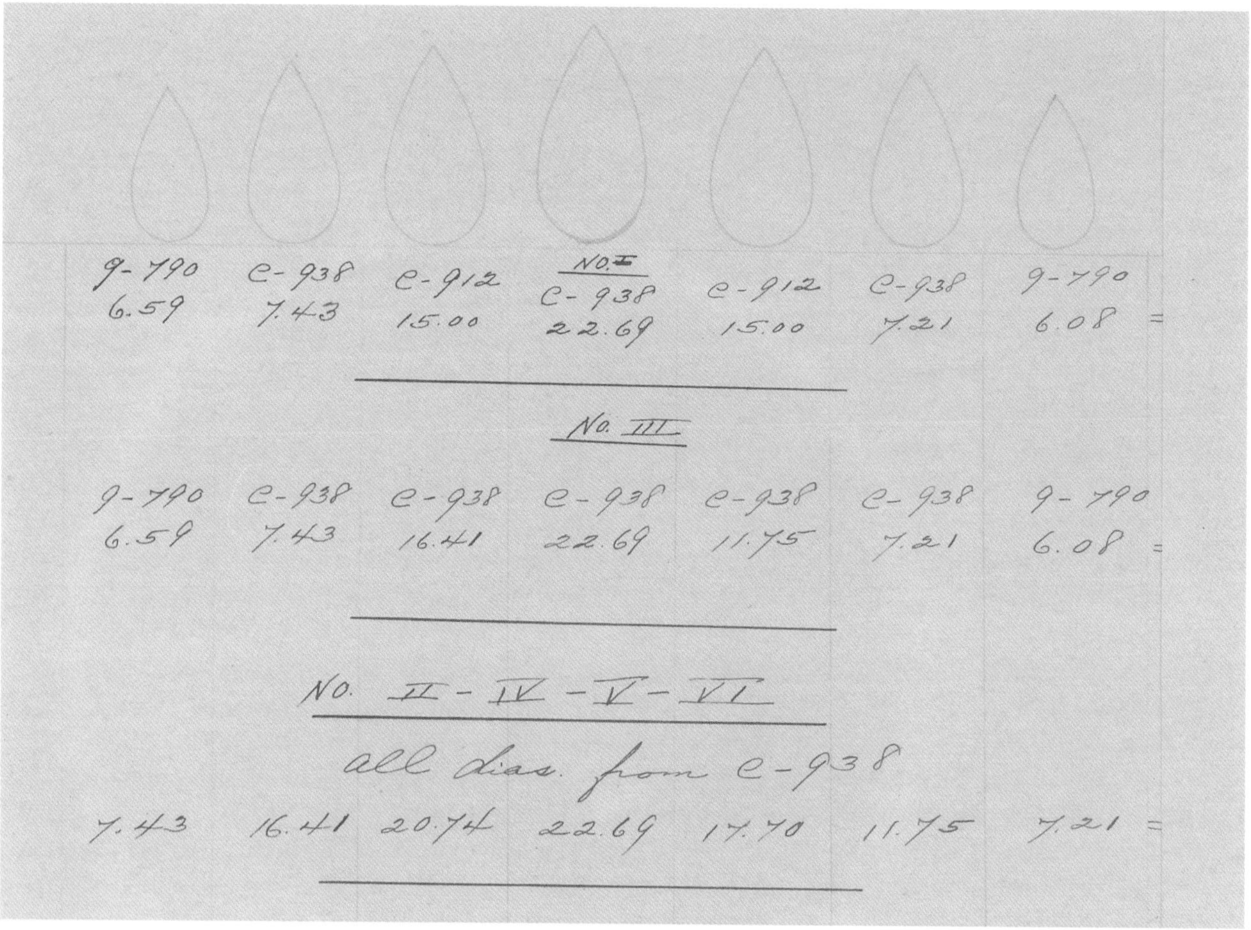

| 9-790 | e-938 | e-912 | No. II
e-938 | e-912 | e-938 | 9-790 |
| 6.59 | 7.43 | 15.00 | 22.69 | 15.00 | 7.21 | 6.08 = |

_____

No. III

| 9-790 | e-938 | e-938 | e-938 | e-938 | e-938 | 9-790 |
| 6.59 | 7.43 | 16.41 | 22.69 | 11.75 | 7.21 | 6.08 = |

_____

NO. II – IV – V – VI

all dias. from e-938

| 7.43 | 16.41 | 20.74 | 22.69 | 17.70 | 11.75 | 7.21 = |

_____

Above: *Design showing alternative ways to set the pear-shaped diamonds of different weights for the necklace shown opposite.*

Opposite: *Necklace in platinum and diamonds. Cartier New York, 1941*
The choker is set with brilliant-cut and baguette-cut diamonds. The two links and the clasp have emerald-cut, baguette-cut, and square-cut diamonds. Seven pear-shaped diamonds, with a total of 78.16 carats, dangle from the choker, connected by diamonds mounted on square chatons.

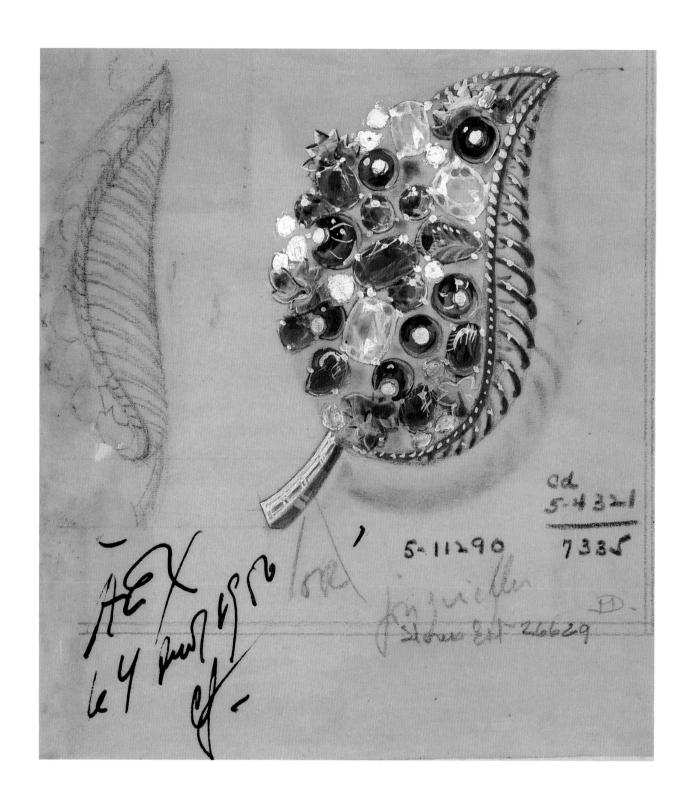

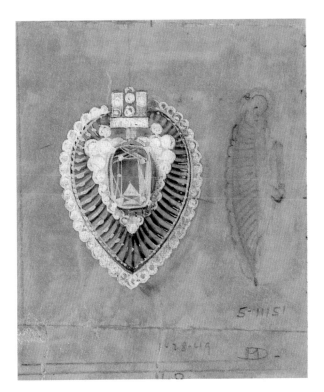

This page and opposite: Original designs for leaf and flower brooches using materials in stock that combine diamond on platinum motifs and gold. Archives Cartier New York, 1940s–50s

*T*he theme of the bird, flying or at rest, tested the skills of Cartier's designers and artisans, who managed to make shapes from hard materials such as gold and platinum resemble live animals. An articulation adroitly placed under the wings emphasized the impression of movement.

Above: *Kingfisher brooch in platinum, gold, emeralds, diamonds, and rubies. Cartier Paris, 1956*
The bird has a pavé diamond body, wings made of engraved leaf-shaped emeralds tipped with diamond and platinum accents, and eyes of oval-cut rubies.

Below left: *Green bird brooch in platinum, gold, emeralds, and diamonds. Cartier Paris, 1942*
A cabochon emerald of 26.9 carats makes up the bird's body. Five more cabochon emeralds adorn one of the tail feathers. Wings, head, and neck are embellished with diamond and platinum motifs.

Below right: *Kingfisher brooch in platinum, gold, emeralds, sapphires, diamonds, and a ruby. Cartier Paris, 1941. Cartier Collection*
The bird has a pavé diamond body and a tail of calibré-cut sapphires. Wings of two large, engraved, leaf-shaped emeralds of 17.66 carats edged with baguette-cut sapphires extend into ten rows of diamonds covered with four motifs of two square-cut sapphires. A cabochon ruby forms the eye and the long back is all gold.

Opposite: *Bird of paradise brooch in platinum, gold, emeralds, sapphires, and diamonds. Cartier Paris, 1944. Cartier Collection*
The gold bird has articulated wings and tail. Four cabochon emeralds compose the body, wings and tail feathers are set in faceted sapphires, and accents are in diamonds and platinum.

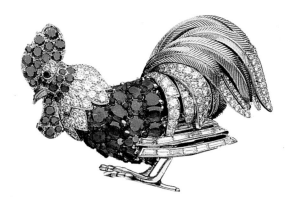

Above: *Rooster brooch in platinum, gold, sapphires, diamonds, and blue enamel. Cartier Paris, 1958*
The head and body are set with sapphires on platinum, with spikes of blue enamel between the sapphires of the body. Brilliant-cut and baguette-cut diamonds on platinum compose the neck and the ends of the wing and tail feathers.

Right: *Bird of paradise brooch in platinum, gold, cabochon sapphires, diamonds, and two engraved cabochon emeralds. Cartier Paris, 1949. Cartier Collection*

Opposite: *Rooster brooch in platinum, gold, rubies, diamonds, and an emerald. Cartier Paris, 1959. Cartier Collection*
The crest and body are set with pear-shaped rubies, the head and tail with brilliant-cut and baguette-cut diamonds on platinum. The eye is a marquise-cut emerald.

*M*ost special orders arise from the desire to have a lasting token, in precious materials, of specific occasions, such as engagements, anniversaries, birthdays. Some of them, however, commemorate extraordinary and sometimes unique events.

Thus, Cartier had the honor of a request to create a powder compact and a cigarette case for the American aviator Jacqueline Cochran. On the gold cases, engraved with maps indicating cities with precious stones, platinum lines marked the itineraries of her historic flights. At the request of the Duke and Duchess of Windsor, Cartier Paris created a souvenir box of a similar type, showing the itinerary of their honeymoon.

Above: *Design for a powder compact whose upper part marks the flight of a bomber, while the lower part shows the first leg of a trip between England and Melbourne, Australia.*

Below: *Design for a cigarette case with the map of the United States and the paths of various flights.*

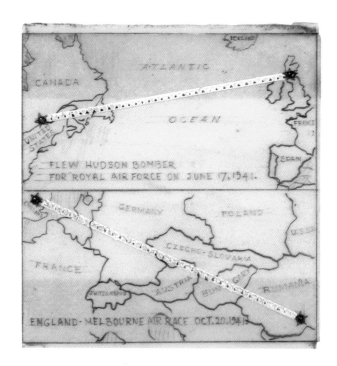

The designs on these two pages all come from Archives Cartier New York, ca. 1937–41.

Opposite: *Design for a charm necklace.*

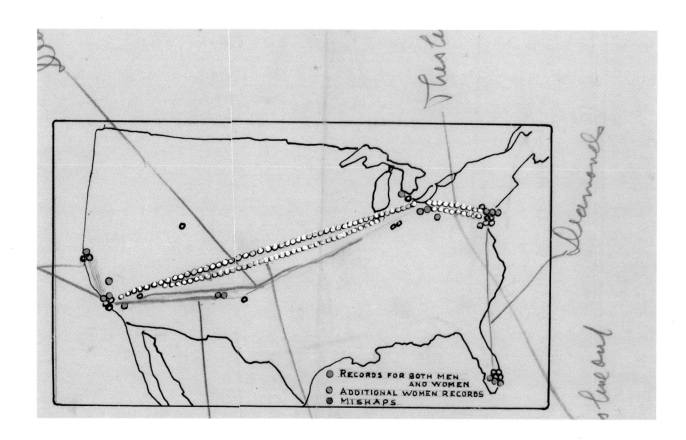

*O*riginal designs for three birds in platinum and gold, set with an amethyst in the center. Archives Cartier New York, 1952–53
These creations of Cartier New York, of smaller dimensions and decidedly less aerial than those of Cartier Paris (see pages 194–95), display a liveliness that suits their size. They call to mind the funny and touching creatures of film animation.

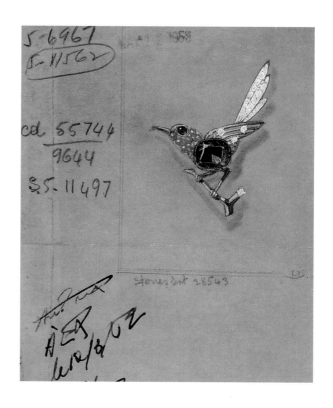

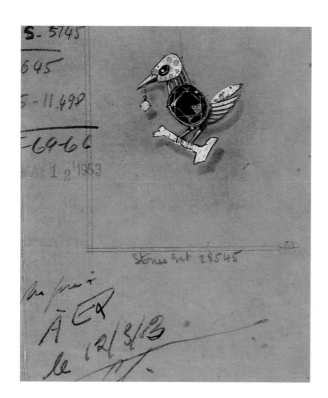

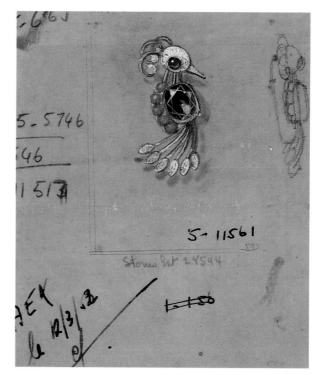

ight: *Dragonfly brooch in platinum, gold, rubies, diamonds, and an emerald. Cartier Paris, 1954. Cartier Collection*
A cabochon emerald forms the insect's body and a square-cut diamond and cailbré-cut rubies the tail. The head is made of diamonds, the eyes of two faceted round rubies, and the two pairs of wings have a tremblant mounting on springs.

Left: *White dragonfly brooch in platinum, diamonds, and sapphires. Cartier New York, 1948 Two pear-shaped diamonds form the body, elongated into a tail of two lines of square-cut diamonds with a tiny brilliant-cut diamond at the tip. Brilliant-cut and baguette-cut diamonds form the wings, and the eyes are two faceted round sapphires. The brooch was made on special order.*

The designs on these two pages all come from Archives Cartier New York

Above: *Ring in platinum, gold, a ribbed emerald bead studded with a diamond, and brilliant-cut and baguette-cut diamonds. Cartier New York, 1951*

Right: *"Tutti frutti" bracelet in platinum, emeralds, sapphires, rubies, and diamonds. Executed ca. 1940 The bracelet is divided into twelve elements, each intersection intended to have an irregular joint that would be invisible in the realized article. It was made on special order.*

Opposite: *"Tutti frutti" brooch in platinum, sapphires, and emeralds. Executed by Cartier New York, 1951 The brooch was made on special order to accompany the bracelet at right.*

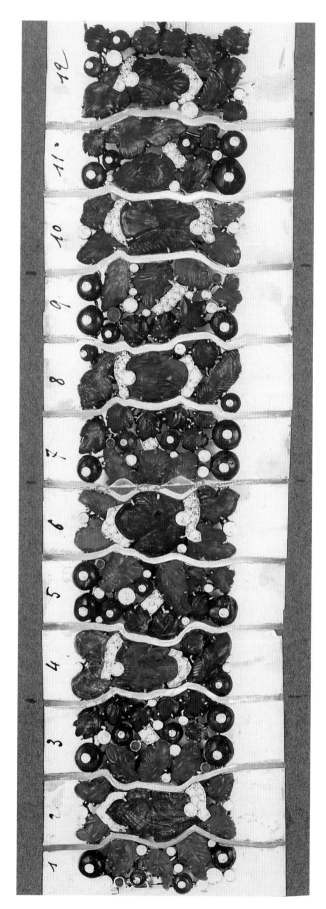

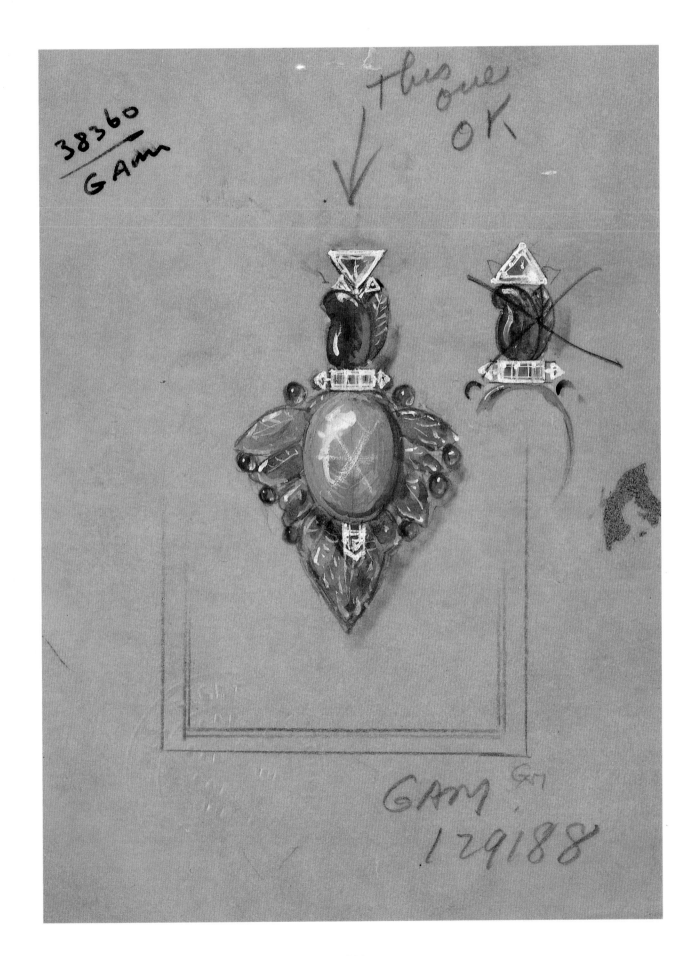

38360
GAm

This one OK

GAm G7
129188

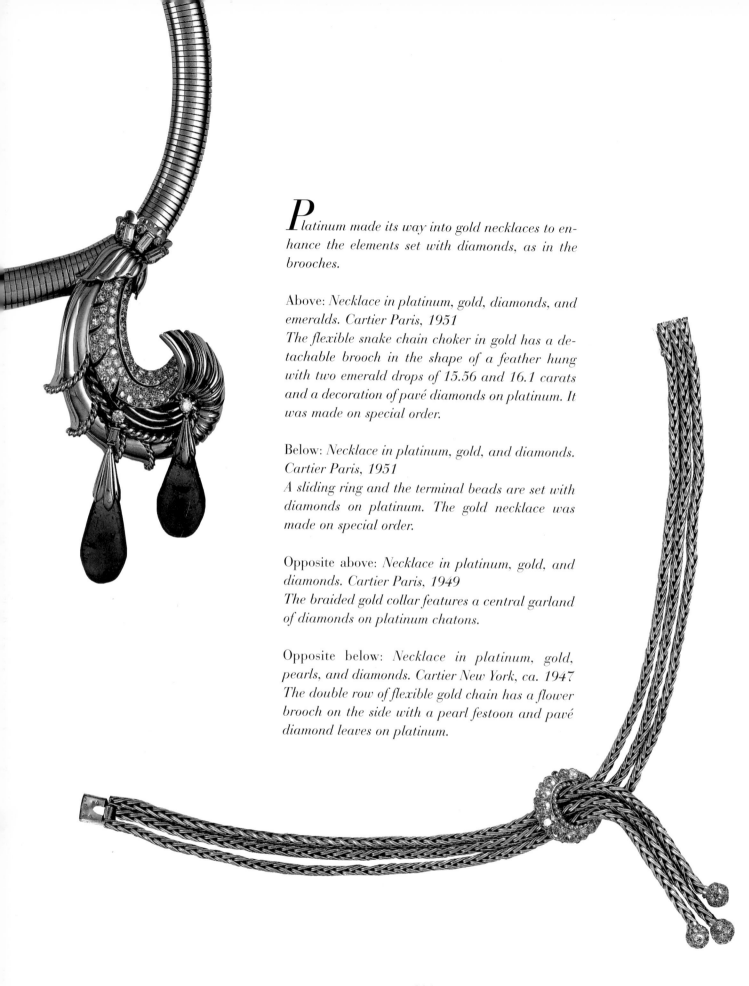

*P*latinum made its way into gold necklaces to enhance the elements set with diamonds, as in the brooches.

Above: *Necklace in platinum, gold, diamonds, and emeralds. Cartier Paris, 1951*
The flexible snake chain choker in gold has a detachable brooch in the shape of a feather hung with two emerald drops of 15.56 and 16.1 carats and a decoration of pavé diamonds on platinum. It was made on special order.

Below: *Necklace in platinum, gold, and diamonds. Cartier Paris, 1951*
A sliding ring and the terminal beads are set with diamonds on platinum. The gold necklace was made on special order.

Opposite above: *Necklace in platinum, gold, and diamonds. Cartier Paris, 1949*
The braided gold collar features a central garland of diamonds on platinum chatons.

Opposite below: *Necklace in platinum, gold, pearls, and diamonds. Cartier New York, ca. 1947*
The double row of flexible gold chain has a flower brooch on the side with a pearl festoon and pavé diamond leaves on platinum.

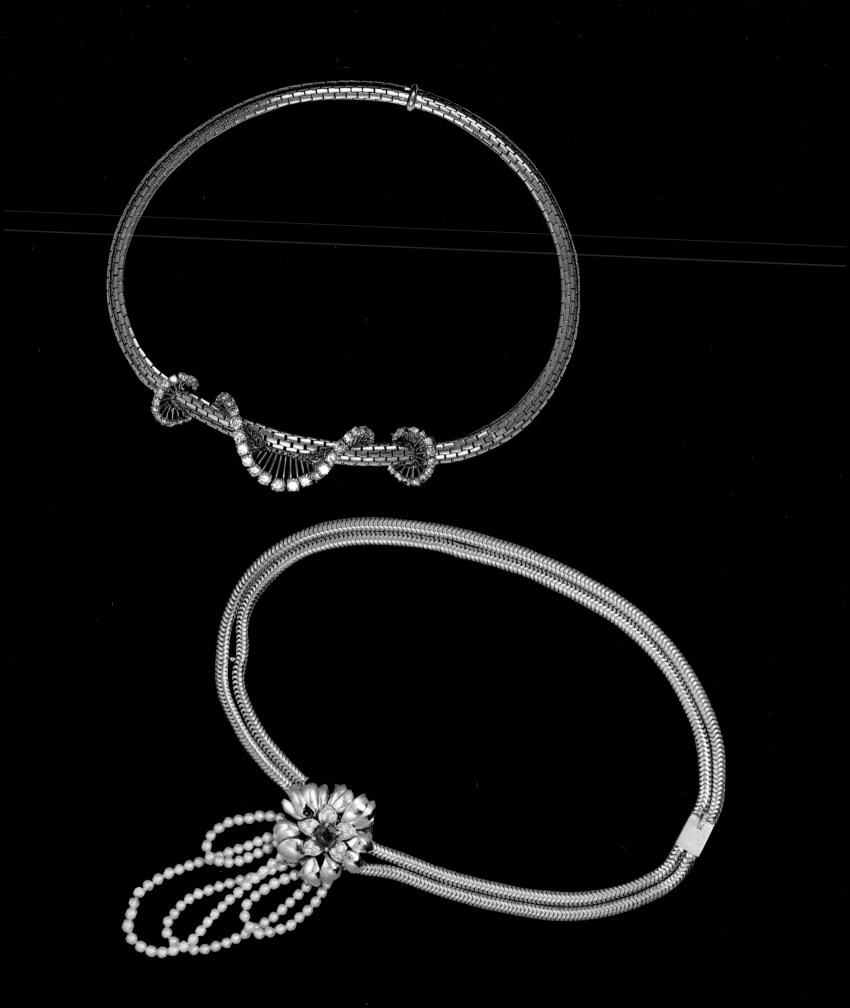

The animal jewels of the 1940s through the 1960s, which might be imaginary (like the chimeras) or naturalistic (like the butterflies or the famous ladybugs), utilized gold and platinum equally.

Right: *Bird brooch in platinum, diamonds, an engraved ruby, and an emerald. Cartier Paris, 1960. Cartier Collection*

Center: *Butterfly brooch in platinum, gold, coral, emeralds, and diamonds. Cartier Paris, 1946 The diamond and platinum butterfly, accented with an emerald, sits on a cabochon coral flower with gold-rimmed, facet-cut emerald leaves and a diamond stem. The brooch originally belonged to the Duchess of Windsor.*

Below left: *Chimera bracelet in platinum, coral, and diamonds. Cartier Paris, 1954. Cartier Collection*

Below right: *Ladybug brooch and earrings in platinum, coral, diamonds, and black lacquer. Cartier Paris, 1958*

Opposite: *Chimera bangle bracelet in platinum, gold, round rubies, emeralds, and diamonds. Cartier Paris, 1960. Cartier Collection*

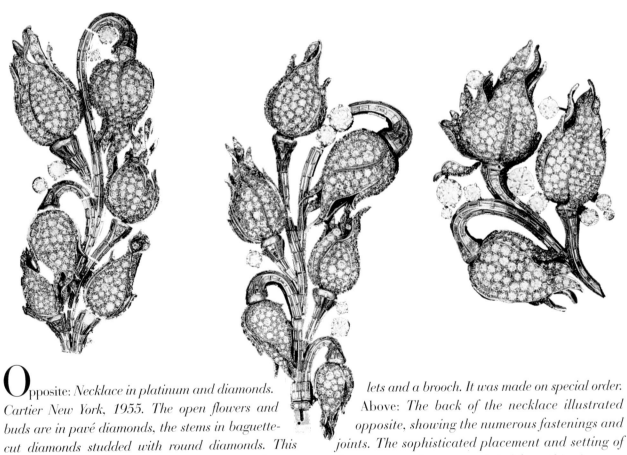

Opposite: *Necklace in platinum and diamonds.
Cartier New York, 1955. The open flowers and
buds are in pavé diamonds, the stems in baguette-
cut diamonds studded with round diamonds. This
three-dimensional necklace detaches into two brace-*lets and a brooch. It was made on special order.
Above: *The back of the necklace illustrated
opposite, showing the numerous fastenings and
joints. The sophisticated placement and setting of
the stones are best appreciated from this view.*

*P*anther brooch in platinum, sapphires, diamonds, and yellow diamonds. Cartier Paris, 1949.
Cartier Collection

A three-dimensional panther of pavé diamonds
studded with sapphires crouches on a cabochon
sapphire weighing 152.35 carats. Marquise-cut
yellow diamonds form the eyes. It originally belonged to the Duchess of Windsor, and at the auction
in Geneva organized by Sotheby's on 2 April 1987,
Cartier bought it back for its collection for
1,540,000 Swiss francs.

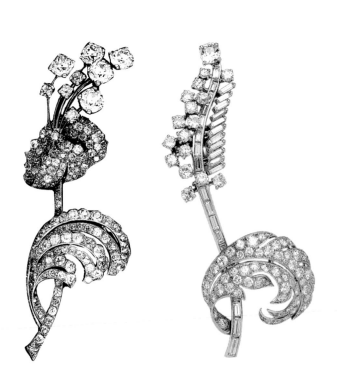

*T*iger parure: brooch and earrings in platinum, gold, diamonds, yellow diamonds, and onyx. Cartier Paris, 1957 (brooch) and 1961 (earrings). Cartier Collection
The paws and tails are articulated. The original owner was Barbara Hutton, the Woolworth heiress.

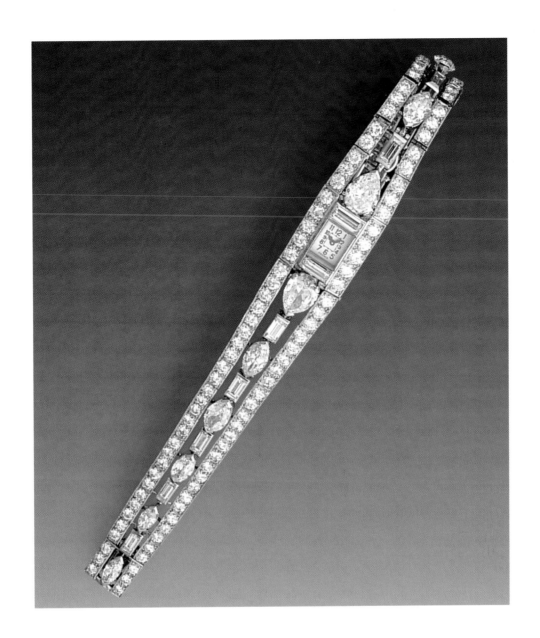

Above: *Bracelet watch for women in platinum and diamonds. Cartier Paris, 1950. Cartier Collection*
Two bands of diamonds enclose a row of pear-shaped, baguette-cut, and navette-cut diamonds. The wind stem is on the back of the case.
Left: *"Baignoire" bracelet watch for women in platinum and diamonds. Cartier Paris, 1967. Cartier Collection*
The bezel is set with two rows of pavé diamonds. The wind stem is on the back of the case.

The stylized palm trees of the 1950s were realized either in gold or in platinum. The latter were entirely set with precious stones, diamonds alone or with colored gems.

Above: *Palm tree brooch in platinum and diamonds. Cartier Paris, 1950*
From four of the brilliant-cut and baguette-cut diamond fronds hang pear-shaped diamonds. The baguette-cut diamond trunk rises from a base of five marquise-cut diamonds.

Below left: *Palm tree brooch in platinum and diamonds. Cartier Paris, 1939*
The articulated trunk is set with brilliant-cut diamonds, the fronds with various shaped diamonds and hung with six briolette diamonds with a total weight of 24.87 carats.

O pposite below right: *Palm tree brooch in platinum, rubies, and diamonds. Cartier Paris, 1957. Cartier Collection*

A cluster of seven Burmese rubies of 23.1 carats, studded with diamonds, bursts from the brilliant-cut and baguette-cut diamond fronds. The articulated trunk is set with brilliant-cut and baguette-cut diamonds. The brooch was made on special order.

Below: *Design for a palm tree brooch. Archives Cartier New York, 1951–52*

The fronds are gold and set with baguette-cut diamonds, with four briolette diamonds hanging from them. The trunk is composed of graduated baguette-cut diamonds.

RÉPUBLIQUE FRANÇAISE.

MINISTÈRE DU COMMERCE ET DE L'INDUSTRIE.

DIRECTION DE LA PROPRIÉTÉ INDUSTRIELLE.

BREVET D'INVENTION.

Gr. 17. — Cl. 5.   **N° 753,508**

Mode de fixation des bijoux sur leur monture.

CARTIER (Société anonyme) résidant en France (Seine).

Demandé le 18 mars 1933, à 14ʰ 49ᵐ, à Paris.
Délivré le 12 août 1933. — Publié le 18 octobre 1933.

La présente invention a pour objet un mode de fixation de bijoux, tels que cabochons, pierres taillées, pendeloques, sur leur monture consistant à ménager à la
5 base de la pierre à sertir, ou dans le feuilletis une gorge dont l'arête intérieure est en retrait par rapport à l'arête extérieure et à sertir dans ladite gorge le bord ou extrémité de la monture, de sorte que cette
10 dernière est absolument invisible. Ce mode de sertissage permet de réaliser un pavage c'est-à-dire une juxtaposition de cabochons ou pierres, dans lequel ces derniers sont jointifs et ne sont pas, comme dans les pa-
15 vages existants, séparés par des bandes métalliques que forment les montures visibles.

Suivant une variante d'exécution destinée plus particulièrement aux pendeloques
20 employées par exemple comme boucles d'oreilles, le mode de fixation consiste à insérer dans la gorge, un anneau ou demi-jonc dont la surface extérieure conique est étamée et à emboîter ce dernier dans un
25 support ou monture à paroi interne conique également étamée, l'assemblage ayant lieu par simple chauffage, assurant la soudure des parties étamées ; on peut ainsi facilement réparer les pendeloques dont l'extré-
30 mité à œil se serait cassé.

La description qui va suivre, en regard du dessin annexé, donné à titre d'exemple, fera bien comprendre de quelle manière l'invention peut être réalisée.

La fig. 1 est une vue de détail à grande 35 échelle du sertissage d'un cabochon, partiellement en coupe.

La fig. 2 est une variante.

La fig. 3 montre un pavage en vue de côté, partie en coupe. 40

La fig. 4 est un plan de la fig. 3.

La fig. 5 montre le sertissage d'une pierre taillée.

Les fig. 6, 7 et 8 se rapportent à la fixation d'une pendeloque sur sa monture. 45

Comme on le voit sur le dessin une gorge a est ménagée à la base du cabochon b (fig. 1 et 2) ou dans le feuilletis de la pierre taillée b^1 (fig. 5), l'arête a^1 intérieure de la gorge étant en retrait par rapport à l'arête 50 extérieure a^2 comme indiqué en h sur la fig. 1. C'est dans cette gorge qu'est serti le bord ou extrémité c^1 de la monture c. Ce bord c^1 peut s'étendre sur tout le pourtour de la gorge (fig. 2) ou former des pattes 55 ou griffes c^1, c^2 serties dans ladite gorge. Grâce au retrait de l'arête a^1 par rapport à l'arête a^2 la monture est absolument invisible ce qui est particulièrement intéressant dans le cas de pavages (fig. 3 et 4). On voit 60 en se reportant à ces figures que les éléments b^2 de pavage (cabochons ou pierres taillées) ne sont plus séparées comme dans les pavages habituels par des bandes métalliques.

Prix du fascicule : 5 francs.

*T*he patent for Cartier's "invisible setting," reproduced opposite, dates back to 1933, but it was
rarely used as it was considered inappropriate for
precious stones, which had to be grooved under the
girdle in order to thread the two sides to hold them
in place.
Above: *Blue rose brooch in platinum, sapphires,
and diamonds. Cartier Paris, 1960. Cartier
Collection*
*The petals are set with calibré-cut sapphires with
the technique known as "invisible setting" (monture invisible) or "mystery setting" (serti mystérieux).
The three leaves are in pavé diamonds with
baguette-cut diamond veins.*

O pposite: *Parakeet brooch in platinum, gold, diamonds, yellow diamonds, and an emerald. Cartier Paris, 1969. Cartier Collection*
The parakeet, perched on a gold branch, is made of yellow diamonds on gold and brilliant-cut and baguette-cut diamonds on platinum. The eye is a round facet-cut emerald.

Above left: *Duckling brooch in platinum, gold, coral, diamonds, and emeralds. Cartier Paris, 1966. Private collection*
A round engraved emerald framed by diamonds forms the body with pavé diamonds used for the head and tail. The eye is a round facet-cut emerald. The beak and feet are in coral.

Above right: *Owl brooch in platinum, gold, coral, sapphires, and diamonds. Cartier Paris, 1963*
Sitting on a branch set with a diamond is an owl of ribbed coral. The head is engraved and the eyes are round facet-cut sapphires. Its horn and collar are in diamonds.

Center left: *Bird brooch in platinum, gold, coral, diamonds, a sapphire, and a pearl. Cartier Paris, 1967*
In a ribbed coral nest rimmed with diamonds on platinum, resting on a branch set with a diamond and a pearl, is a small bird with a head of pavé diamonds on platinum and an eye of round facet-cut sapphire. Its beak and wings are gold.

Center right: *Bird brooch in platinum, gold, pearls, coral, and a diamond. Cartier Paris, 1971*
The two small birds have baroque pearl bodies on gold. They sit on a branch of coral set with a diamond.

Below: *Tortoise brooch in platinum, gold, a cushion-cut diamond surrounded by citrines, turquoises, and brilliant-cut diamonds. Cartier Paris, 1961*

Above: *Flower brooch in platinum and diamonds. Cartier Paris, 1962*
The petals are pavé set with diamonds the natural color of bright yellow, the stem baguette-cut diamonds, and the leaf pavé diamonds.

Center right: *Flower brooch in platinum and diamonds. Cartier Paris, 1965*
Each of the three white flowers is composed of a group of six marquise-cut diamonds, supported by stems in baguette-cut diamonds.

Center left: *Camellia brooch in platinum, gold, turquoises, and diamonds, Cartier Paris, 1967. Cartier Collection*
The petals are in turquoises and marquise-cut diamonds on platinum, edged in engraved gold.

Below: *Coral rose brooch in platinum set with a diamond in the center and with a pavé diamond leaf. Cartier Paris, 1962*

Opposite: *Flower brooch in platinum, gold, emeralds, and diamonds. Cartier Paris, 1964*
This stylized flower has irregular-shaped leaves or petals bordered in brilliant-cut diamonds with wavy lines of baguette-cut diamonds inside. The center is decorated with six facet-cut emeralds set in gold and five marquise-cut diamonds. The branch is set with baguette-cut diamonds.

THE NEW ERA
From 1968 Toward the Year 2000

With the successful adventure of Les Must, Cartier renewed itself. As the 1990s began, platinum came to the fore not only in the area of fine jewelry but also in its new line of New Jewelry and, especially, of watches.

The main entrance of the Centre Culturel Georges Pompidou in Paris (architects: Renzo Piano and Richard Rogers, 1970s).
The Centre Culturel Georges Pompidou admirably illustrates the high tech culture of the 1970s.
The metallic substructure is open to view and provides the building's aesthetic dimension as well.

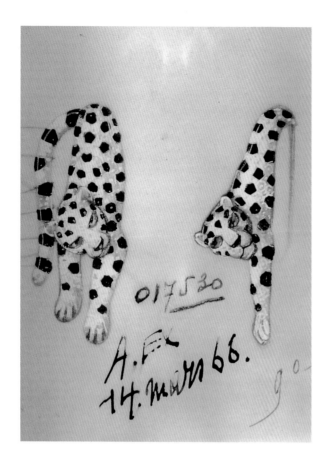

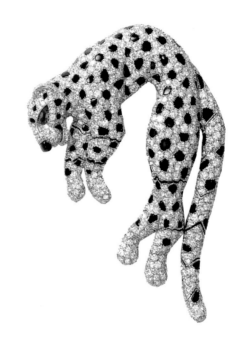

Opposite right: *Panther brooch in platinum, diamonds, onyx, and emeralds. Cartier Paris, 1966 The recumbent panther, with pear-shaped facet-cut emerald eyes, has spots of onyx on a pavé diamond coat. Its tail is made of alternating pavé diamond rondelles and onyx tubes. Head, paws, and tails are articulated.*

Above left: *Design for a panther brooch, face and profile. Archives Cartier Paris, 1966*

Above right: *Leopard brooch in platinum, diamonds, onyx, and an emerald. Cartier Paris, 1970 The body is pavé diamonds with irregular-shaped onyx, the eye a pear-shaped facet-cut emerald, the nose in onyx. The paws and tail are articulated.*

Right: *Panther brooch in platinum, diamonds, sapphires, onyx, and emeralds. Cartier Paris, 1989 The pavé diamond body is spotted with irregular-shaped sapphires. The head, seen full face, is movable, with eyes in pear-shaped facet-cut emeralds and onyx nose. The paws and tail are articulated.*

Opposite left: *Panther bracelet in platinum, diamonds, onyx, and emeralds. Cartier Paris, 1952. Private collection The stretched-out cat is spotted with irregular-shaped onyx on its pavé diamond coat and has two pear-shaped emerald eyes.*

*A*ll of these wristwatches are made of platinum, the dials with burgundy-colored Roman numerals and the wind stems set with a cabochon ruby.

Clockwise from upper left: "Santos-Dumont" for

men, "Ellipse" for women, "Ceinture" for men, and "Baignoire" for women.

Opposite, left to right: "Tortue," "Tank L. C.," and "Vendôme L. C." for women. Cartier Paris, 1979

*L*eft, above and below: *Bracelet and rings in three and seven bands. Cartier, 1985*
These are all-platinum versions of the Rolling Rings in three golds, whose famous "forebear" was made in 1924 by Louis Cartier for the poet Jean Cocteau. These particular versions were made on special order.

*O*pposite: *Bracelet, rings, and earrings in platinum. Cartier, 1985*
The basic structure is flat, ribbed platinum in a curve that is rounded off at either end with the Cartier initial C.

The "Rhinocéros" line was born in 1988, when Cartier decided to join the fight led by the World Wildlife Fund, or WWF, to save this species threatened with extinction. Cartier offered many variations in platinum jewelry.

Above: *Lapel brooch in platinum*

Below: *Bracelet with three rhinoceros charms in platinum*

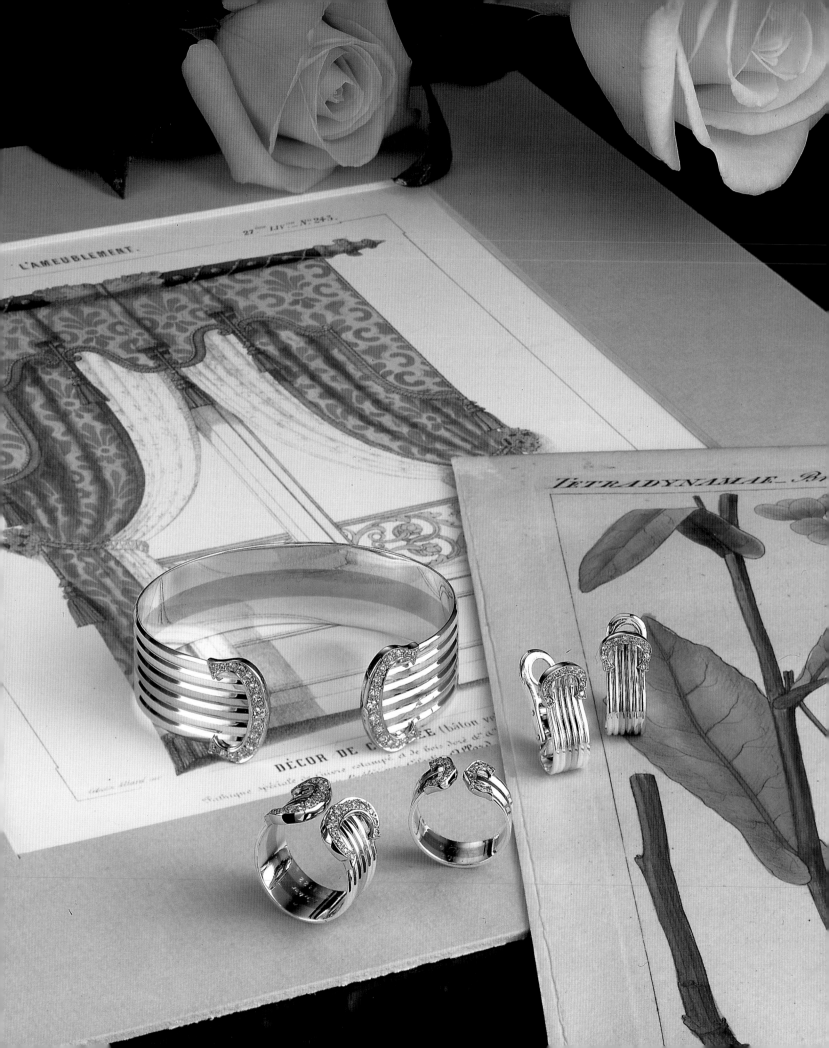

Above: *Design for the "Bleuet" necklace.*
In the "Bleuet," or cornflower, necklace, a simple strand of diamonds circles the neck, enlarging into a double central motif of alternating oval sapphires and navette-cut diamonds framed by brilliant-cut diamonds. The links on the side and the center are made of baguette-cut diamonds.

Below: *"Bleuet" bracelet in platinum, sapphires, and diamonds. Cartier, 1986*
A row of diamond chatons enlarge in a double central motif of oval sapphires alternating with navette-cut diamonds, the motif entirely surrounded by brilliant-cut diamonds. The links on the side and the center are made of baguette-cut diamonds.

Opposite: *"Tamara" necklace in platinum, diamonds, pearls, and a blue diamond. Cartier Paris, 1985*
The necklace is composed of geometric links in platinum and diamonds, recalling the Art Deco style, with a vertical central motif. The double pairing of two white pearls weighing about 48.8 grains on the necklace and two gray pearls weighing 108.17 grains on the "tie" creates an unusual and elegant effect. The central motif suspends a pear-shaped blue diamond of 5.04 carats.

Above: *"Egg and Panther" table clock in platinum, gold, aventurine, diamonds, and sapphires. Cartier Paris, 1987*
The diamond and gold clock is hidden inside the egg, which is covered entirely with 602 invisibly set square-cut and calibré-cut sapphires. The panther draped on top of the egg is in platinum set with diamonds and sapphires. These elements are presented on an aventurine cushion trimmed with a band of diamonds on platinum and embellished with four tassels in platinum, diamonds, and sapphire beads. The base is gold.

Above: *"Tortue Rivière" bracelet watch in platinum and diamonds. Cartier Paris, 1982*

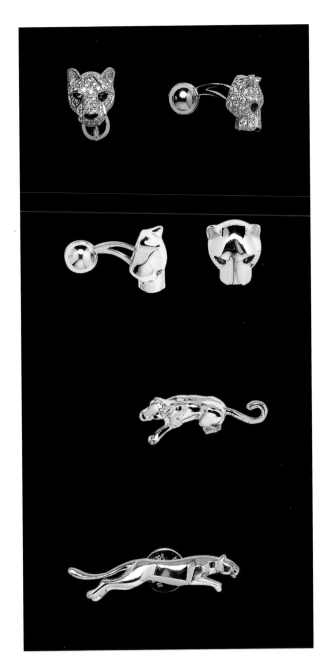

Above: *"Tête de Panthère" watch-jewel, open and closed, in platinum, diamonds, onyx, and emeralds. Cartier Paris, 1984*
The panther bracelet has links of pavé diamonds and rectangular onyx accents on platinum. The dial is hidden under a panther head in pavé diamonds and onyx, with emerald eyes.

Right: *Panther jewels for men. Cartier Paris, 1986*
"Tête de Panthère" cuff links in platinum come in a pavé diamond version (top) or without diamonds (second from top). The nose is onyx, the eyes emerald. Two lapel brooches in platinum: "Mowgly" (third from top) and "leaping panther" (below).

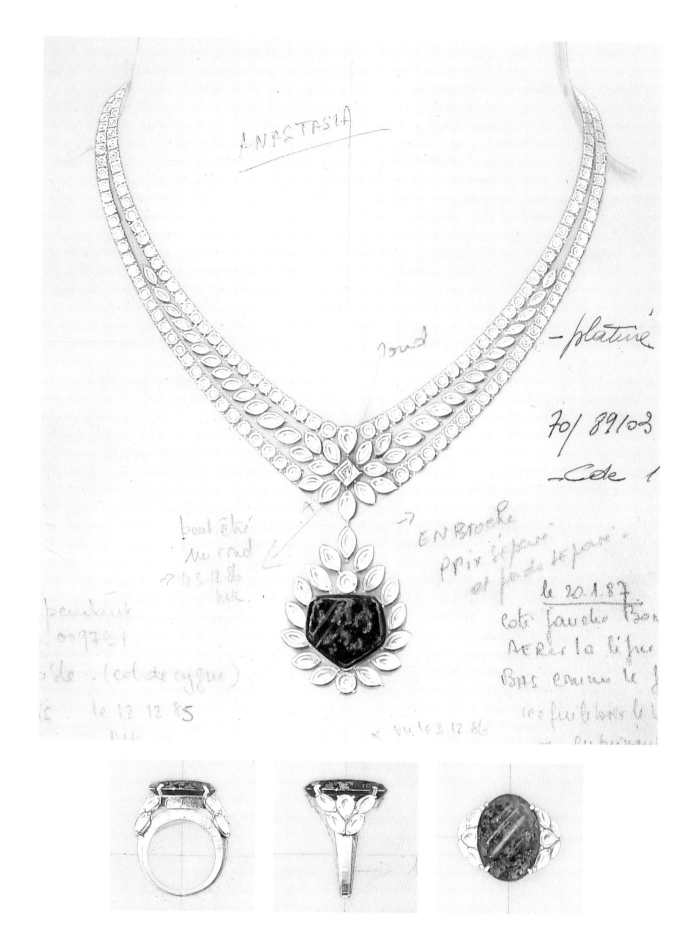

234

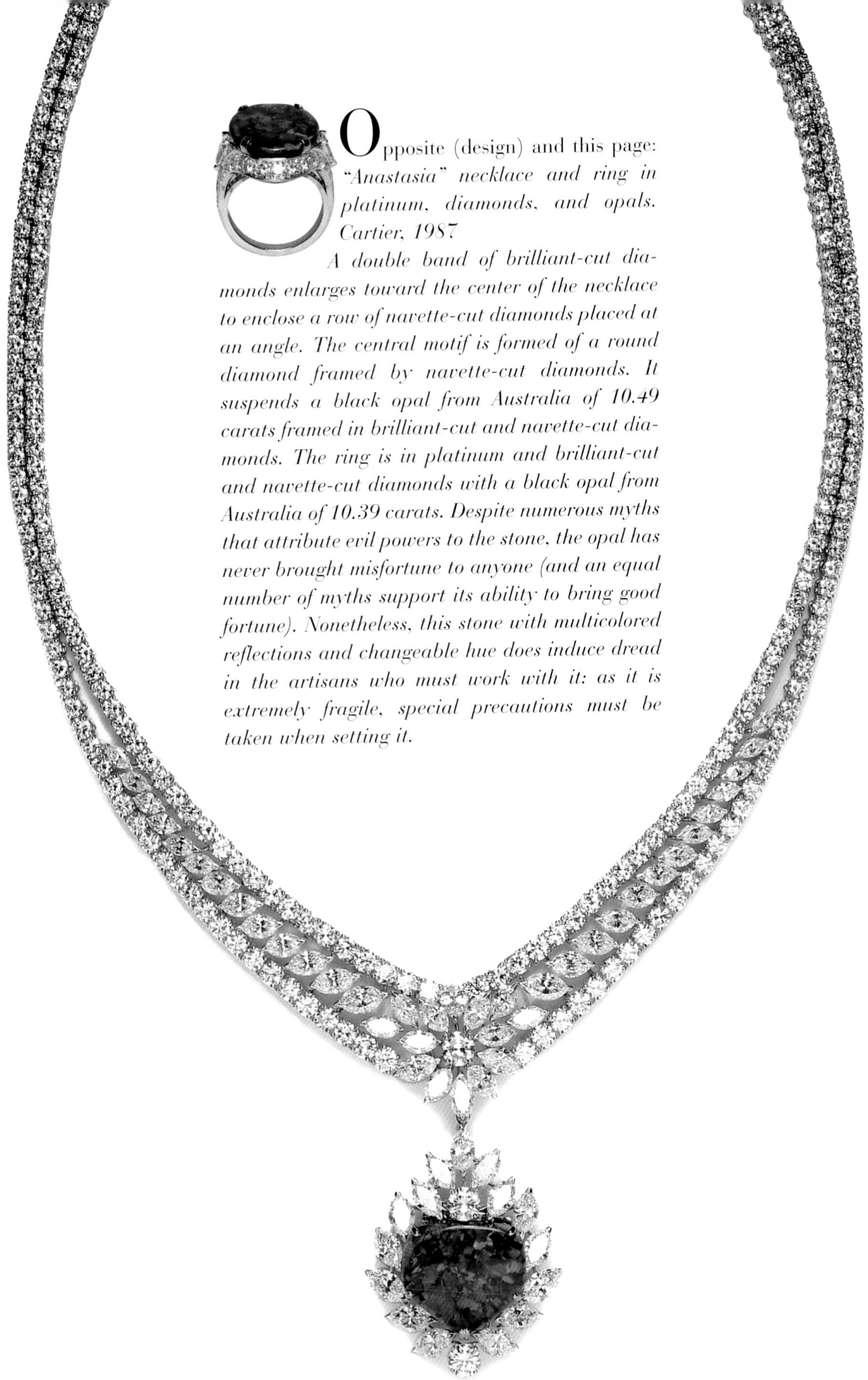

Opposite (design) and this page: "*Anastasia*" necklace and ring in platinum, diamonds, and opals. Cartier, 1987

A double band of brilliant-cut diamonds enlarges toward the center of the necklace to enclose a row of navette-cut diamonds placed at an angle. The central motif is formed of a round diamond framed by navette-cut diamonds. It suspends a black opal from Australia of 10.49 carats framed in brilliant-cut and navette-cut diamonds. The ring is in platinum and brilliant-cut and navette-cut diamonds with a black opal from Australia of 10.39 carats. Despite numerous myths that attribute evil powers to the stone, the opal has never brought misfortune to anyone (and an equal number of myths support its ability to bring good fortune). Nonetheless, this stone with multicolored reflections and changeable hue does induce dread in the artisans who must work with it: as it is extremely fragile, special precautions must be taken when setting it.

"Léopoldine" necklace and earrings in platinum, diamonds, and sapphires. Cartier, 1987

The necklace is in platinum and diamonds in a millegrain setting. The diamond links are bordered top and bottom with a laurel leaf pattern set with diamonds. Three large decorative motifs feature a briolette cabochon sapphire framed in laurel leaves and a row of diamonds. In the earrings, a fall of diamonds on platinum dangles from the briolette cabochon sapphire.

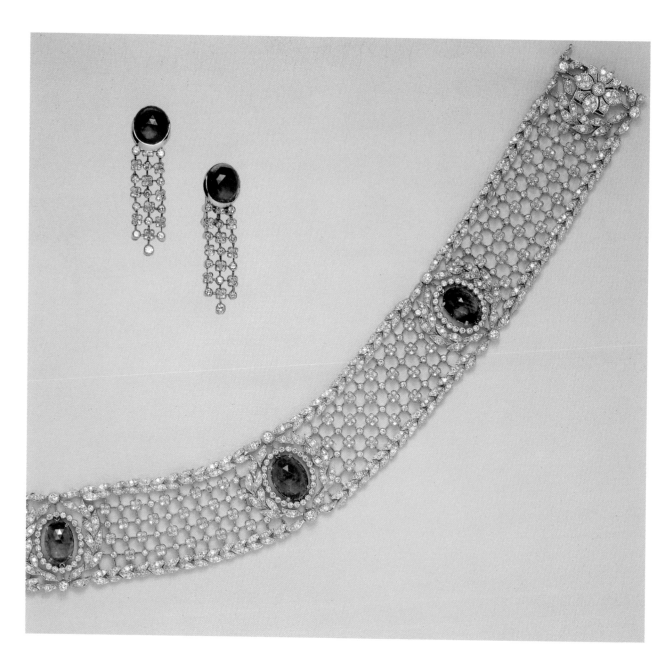

*B*elow and right: *"Oliban" necklace and ear-rings in platinum and diamonds. Cartier, 1990*
A mesh of diamond collets on platinum is bordered top and bottom with a line of diamonds. The ear-rings are also made of a mesh of diamond collets on platinum.

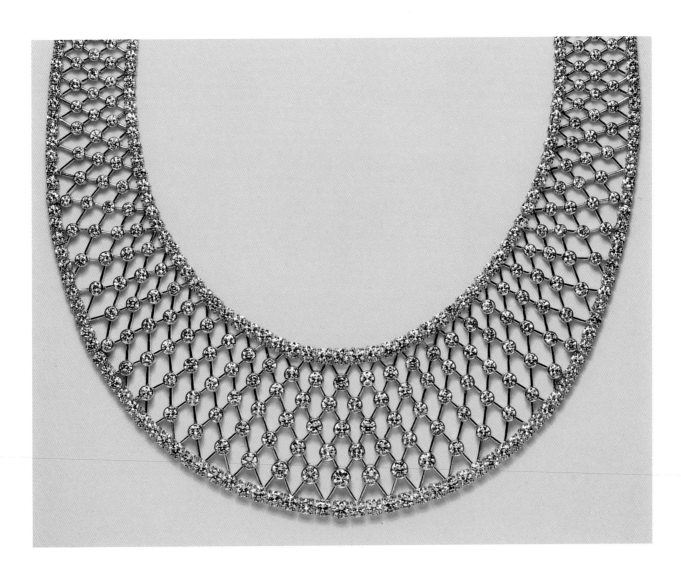

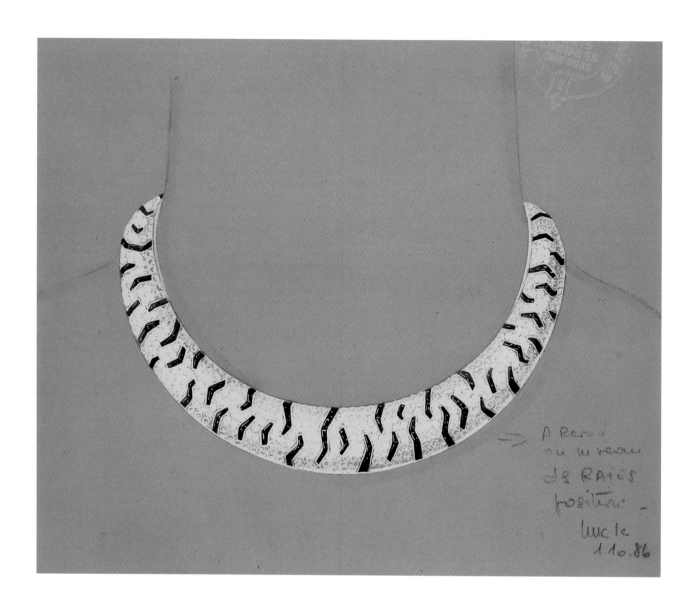

*C*artier's version of tachisme *(the postwar French art movement named for the blot, stain, spot, or drip of its technique), specifically, the abstract representation of the spotted coat of the large cats, appeared in 1914 on a watch set with round cabochon onyx. In its most contemporary expression, it employs the tallow-cut cabochon, which permits a more realistic effect, in order to capture the irregular pattern of the tiger's stripes and the panther's markings.*

Above (design) and opposite above: *Tiger necklace in* platinum, diamonds, and onyx. Cartier Paris, 1992 *Articulated elements in platinum are pavé set with round diamonds and shaped onyx that describes the tiger's stripes.*

Opposite below: *"Sierra" bracelet watch in plati-num, diamonds, and onyx. Cartier Paris, 1989 The bracelet is formed of a triple band of diamonds with onyx and diamond borders. The rectangular watch with pavé diamond dial fits seamlessly in the bracelet.*

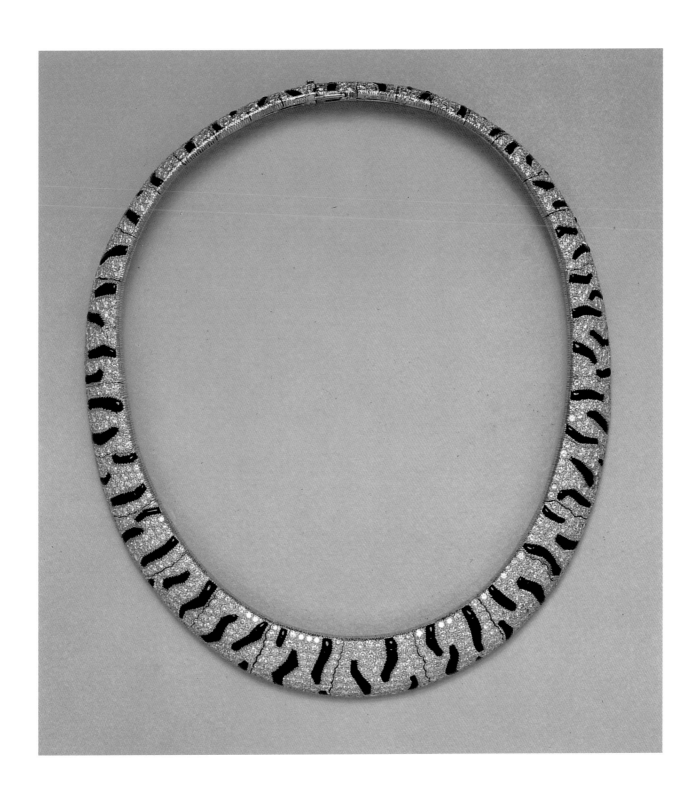

This page and opposite: *"Mayerling" parure in platinum and diamonds. Cartier Paris, 1990 Earrings (above) have a double flower motif, the petals in platinum and pavé diamonds and pistil in four old mine-cut diamonds. The bracelet (below) gives a lacy effect with three rows of three- and four-lobed motifs between two bands of pear-shaped diamonds, enhanced with three flower motifs, each with six pavé diamond petals and three old mine-cut diamonds as pistils.*

Opposite, top to bottom: *Necklace, earrings, and bracelet watch form a parure, all realized in flower motifs with pavé diamond petals and old mine-cut diamonds as pistils.*

O*pposite: "Panther Rivière" watch, in plat-inum, diamonds, and sapphires. Cartier, 1986*
The bracelet is composed of three rows of baguette-cut diamonds bordered on either side with baguette-cut sapphires. The panther hides the dial.

Top: *"Bettina" ring in platinum and diamonds (three views). Cartier Paris, 1993*
The cushion-cut diamond of 12.93 carats is set off by brilliant-cut and baguette-cut diamonds.

Above, clockwise from upper left: *"Eve" ring in*

platinum and diamonds. Cartier, 1990
An emerald-cut diamond of 6.47 carats is enhanced by brilliant-cut and baguette-cut diamonds. The ring was made on special order.
"Deux têtes croisées" (two crossed heads) ring in platinum and diamonds. Cartier, 1991
Diamonds on platinum set off two perfect pear-shaped diamonds (D, IF) of 6.03 and 6.36 carats.
"Delphine" ring in platinum, diamonds, and a sap-phire. Cartier, 1993
The Kashmiri cushion-cut sapphire weighs 10.26 carats.
"Lisa" ring in platinum, diamonds, and a sap-phire. Cartier, 1991
The Burmese sapphire weighs 14.07 carats and is surrounded by brilliant-cut and pear-shaped diamonds.

Above left and opposite (design): *"Duetta"
brooch in platinum, gold, emeralds, diamonds,
and yellow diamonds. Cartier, 1991*
*The brooch is composed of two intertwined flowers
—one in gold with multiple five-petaled motifs,
pistils in the shape of round emeralds, and stem
and leaves in pavé diamonds, the other in the form
of a daisy with diamond on platinum petals and
pistil pavé-set with yellow diamonds on gold.*

Above right and right (design): *"Neige" brooch in
platinum and pavé diamonds. Cartier, 1991*

DUETTA

In 1991, Cartier revived its tradition of the precious good-luck charm. In gold, white gold, or platinum, alone or set with diamonds, these charms illustrate some of the symbols and creations that hold particular meaning for Cartier in its century and a half of history: the number 13, its address on the rue de la Paix; the airplane, in homage to the exploits of Alberto Santos-Dumont; the panther, the company's mascot; the flamingo, beloved of the Duke of Windsor; the elephant, hero of the latest creations in Cartier's collection "Sur la route des Indes" (On the road to the East Indies). At the same time, watches were greatly reduced in size to become miniature masterpieces.

Below left: *Designs for two charms, one in the form of a watch, the other a ladybug, in platinum and diamonds. Paris 1990*

Below right: *"Snoopy" charm, made in honor of Charles Schultz, the creator of the cartoon "Peanuts." Cartier New York, 1972*

Opposite: *Charm bracelet and charms in platinum and diamonds. Cartier, 1991*
The bracelet is in platinum, the number 13, elephant, panther, and flamingo charms in pavé diamonds on platinum. The airplane is in platinum with a square-cut diamond.

This page shows various recent versions of the animal theme.

Above: *"Jumbo" necklace in platinum, diamonds, and rubies. Cartier, 1992*
The small and large pavé diamond elephants, with ruby eyes, are articulated.

Below: *"Satki" jewels for men: cuff links in platinum and lapel brooch in platinum and diamonds*

Right: *"Verouchka" brooch in platinum, diamonds, and a pearl. Cartier, 1990*
The crested crane is in pavé diamonds on platinum, the crest made of platinum wires. The bird perches on a slightly baroque pear-shaped pearl of 99.7 grains.

"*C*onstellation" necklace in platinum and diamonds. Cartier, 1991
The flexible chain is made of four strands of elongated hexagonal motifs set with brilliant-cut diamonds on platinum alternating with baguette-cut diamonds. The two ends of the chain finish in a half-moon motif of brilliant-cut and baguette-cut diamonds, from which hang two square cords made of cubes, each side faced with a diamond. The two rounded motifs midway down the cord and the pear-shaped balls that end it are pavé-set with various-sized diamonds.

Above left: *"Panthère" necklace in platinum, diamonds, and onyx. Cartier, 1992*
The articulated elements in pavé diamonds and irregular-shaped cabochon onyx, imitating the coat of the big cat, increase in size as they approach the center.

Above right: *"Balum" bracelet in platinum, diamonds, and sapphires. Cartier, 1991*
Two panthers in pavé diamonds and cabochon sapphires on platinum face each other over an oval cabochon sapphire on a pavé diamond base.

Right: *Design for "Ida" ring. Cartier, 1991*
The panther heads in pavé diamonds and sapphires with emerald eyes gnaw on a round cabochon sapphire on a platinum base. The body of the ring is in polished platinum.

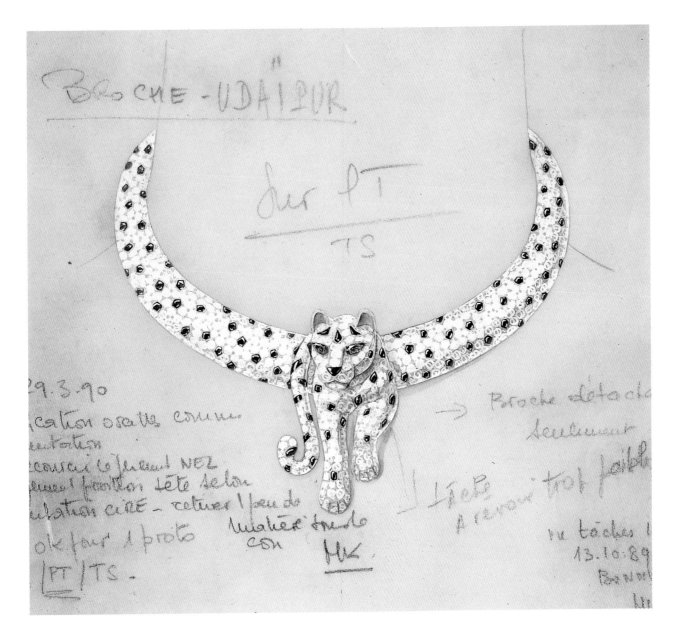

Above: *Design for "Udaïpur" necklace. Cartier, 1991*
To an articulated platinum necklace set with pavé diamonds and cabochon sapphires is attached a detachable panther clip brooch with pavé diamond and sapphire coat and emerald eyes.

Below: *"Panthère" lapel brooch and cuff links in platinum, with emerald eyes.*

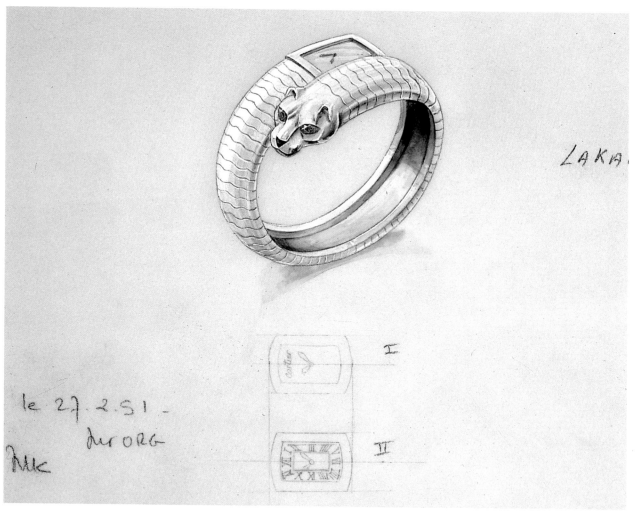

LAKA

le 27.2.51 —
JurORG
MK

I

II

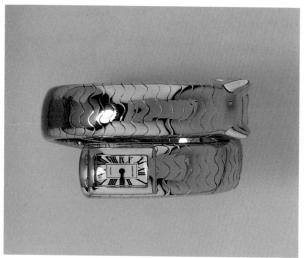

Above left and right and top (design): "Lakarda" bracelet watch in platinum, ebony, and emeralds. Cartier Paris, 1991

In this elegant naturalistic sculpture, the cat's head is set with ebony and two emerald eyes. The flexible springy bracelet goes on the arm without a clasp.

"Anthalia" necklace in platinum and diamonds.
Cartier Paris, 1994
This flexible collar is composed of a succession of geometric column motifs set with 453 diamonds (with a total of more than 85 carats), that graduate in size toward the center of the necklace.

253

*R*ight: *"Taïmango" bracelet watch in platinum, diamonds, onyx, and emeralds. Cartier, 1992*
This flexible spring bracelet incorporates a watch in platinum set entirely with diamonds. The watch is hidden beneath the platinum and pavé diamond panther with an onyx nose and emerald eyes.

Below left and right: *"Tatiana" bracelet in platinum, diamonds, onyx, and emeralds. Cartier, 1992*
The spiraling flexible spring bracelet in platinum and pavé diamonds ends in a stretched-out tiger in platinum and diamonds, with onyx stripes and emerald eyes.

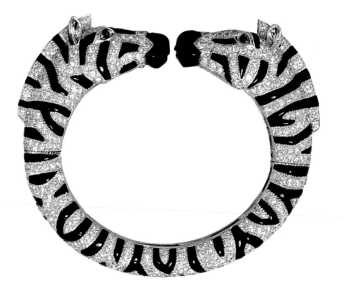

Left: *"Horus" lapel brooch in platinum. Cartier 1991*
This miniature sculpture updates a traditional Egyptian theme.

Above right (design) and center: *"Ricky" brooch in platinum, diamonds, onyx, gold, yellow diamonds, and an emerald. Cartier, 1990*
The raccoon is in platinum and diamonds with onyx stripes and eyes of yellow diamond set in onyx. It holds in its gold forepaws a Colombian cabochon emerald. (The design called for a cabochon of cat's-eye chrysoberyl.)

Above right: *"Bora Bora" brooch in platinum, diamonds, sapphires, onyx, and rock crystal. Cartier, 1990*
A pair of lovebirds in platinum and diamonds, with tails of two rows of baguette-cut sapphires and onyx beaks, sit on a perch made of rock crystal.

Below: *"Bantou" bracelet in platinum, diamonds, onyx, and emeralds. Cartier, 1992*
The two zebra heads pivot on hinges to open the bangle bracelet in platinum and pavé diamonds with onyx stripes. Their muzzles are in onyx and their eyes round emeralds.

"*O*rchidée" *parure in platinum, diamonds, and sapphires. Cartier, 1991*
The necklace (above, design) has three strands of round diamonds with a removable orchid clip brooch in the center. Its petals are set with diamonds and calibré-cut sapphires on platinum. The pistil is a cascade of seven threads of round diamonds ending in ten pear-shaped diamonds. A miniclip brooch (above right) is in platinum, diamonds, and calibré-cut sapphires. The earrings (right) resemble the detachable clip brooch of the necklace.

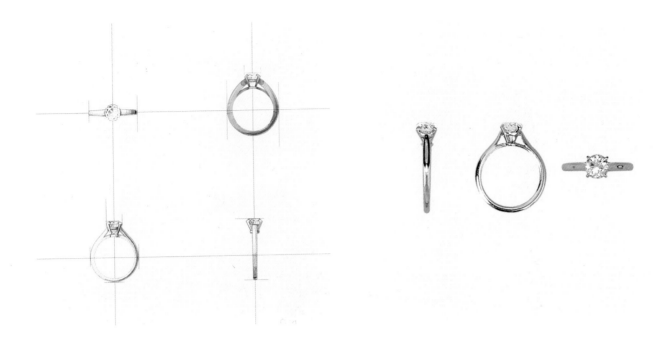

Above left (design) and right: *Engagement ring in platinum and a diamond. Cartier, 1993*
The earliest solitaire settings appeared in the last decade of the nineteenth century. At that time, the chaton and the shank of the ring were fashioned in the Garland style. Through the years, the lines of this setting became progressively simplified until in 1993, with the "MK" ring, it embodied the jewelers' art in this particular realm. The "MK" ring wed the lightness of the four claws in platinum, which held the stone with the least constraint, to the elegance and roundness of the shank, made of a simple platinum hoop. The discretion of the setting and the whiteness of the metal direct attention entirely on the diamond's fire.

Right: *Design for a Louis Cartier "C" solitaire ring. Cartier, 1995*
On either side of the band the chaton outlines the Cartier C.

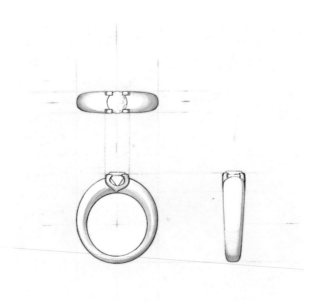

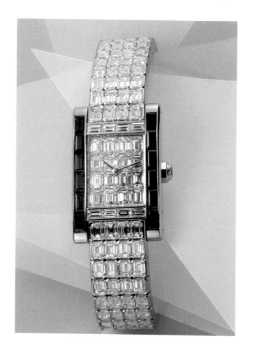

This page and next: *Several jeweled watches from the 1990s*
Clockwise from upper left: *"Tank Rivière" in platinum, sapphires, and diamonds. Cartier, 1992; "Pasha Rivière" skeleton watch in platinum, diamonds, and cabochon sapphires. Cartier, 1993;*

"Vendôme Tutti Frutti" in platinum, diamonds, and sapphires. Cartier, 1993; and "Baignoire Rivière" in platinum and diamonds. Cartier, 1994

Opposite: *"Mini-Tank Rivière" in platinum and diamonds. Cartier, 1994*

258

Above left: *"Tank L. C."* wristwatch, medium model in platinum, silver, and a sapphire. Cartier, 1992
The dial is in silver and the wind stem is set with a cabochon sapphire. The watch has a mechanical movement with manual winding.

Above right: *"Vendôme"* wristwatch in platinum, silver, and a sapphire. Cartier, 1994
The dial is in silver and the wind stem is set with a cabochon sapphire. The watch has a mechanical movement with manual winding.

Above center: *"Diabolo"* wristwatch for men in platinum and diamonds. Cartier, 1993
The wind stem and yokes are set with five brilliant-cut diamonds. The watch has a mechanical movement with manual winding.

Opposite clockwise from the top: *"Pasha"* wristwatch in platinum. Cartier, 1992
The watch has an automatic mechanical movement with minute repeater and cathedral striker.
"Diabolo" wristwach for men in platinum and sapphires. Cartier, 1993
The wind stem and yokes are set with five cabochon sapphires. Automatic mechanical movement with power reserve, seconds counter, and skeleton back.
"Tank Américaine" wristwatch in platinum, silver, and a sapphire. Cartier, 1994
The dial is in silver and the octagonal wind stem is set with a sapphire. The watch has a mechanical movement with manual winding.
"Tonneau" wristwatch for men in platinum, silver, and a sapphire. Cartier, 1992
The dial is in silver and the wind stem is set with a cabochon sapphire. The watch has a mechanical movement with manual winding.

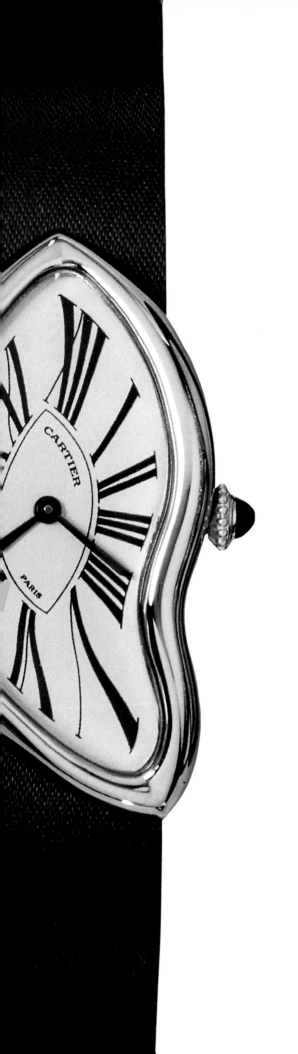

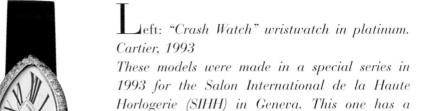

Left: *"Crash Watch" wristwatch in platinum. Cartier, 1993*
These models were made in a special series in 1993 for the Salon International de la Haute Horlogerie (SIHH) in Geneva. This one has a mechanical movement with manual winding.

Below: *"Santos-Dumont" wristwatch for men in platinum and pink gold. Cartier, 1994*
This model was made in a limited edition of ninety to celebrate the ninetieth anniversary of the first "Santos-Dumont" watch. The dial is in pink gold, and the Roman numerals were hand-painted on the dial. The watch has a mechanical movement with manual winding.

Opposite above and below left: *Skeleton watch in platinum. Cartier, 1993*
The mechanical movement with manual winding is a true piece of sculpture.

Opposite below right: *Pocket watch in platinum and sapphires. Cartier Paris, 1993*
The watch, with a mechanical movement, is set with fifty-two baguette-cut sapphires. The dial gives double hour numerals.

Above left: *Design for a "Tank Chinoise" wristwatch in platinum, presented at SIHH, Geneva, in 1995. The watch has a mechanical movement with manual winding.*

Above right: *Design for a "Tortue Chinoise" wristwatch in platinum, presented at SIHH, Geneva, in 1995. The watch has a mechanical movement with manual winding.*

Below, left to right: *Two square watches with bullet-shaped lugs in platinum and diamonds. Cartier, 1995 The two models, one with a leather strap, the other with a platinum bracelet, have a case set with 157 diamonds. They were made on special order.*
Design for a "Tonneau" wristwatch in platinum and diamonds, presented at SIHH, Geneva, in 1995. This model is set with 48 diamonds. It has a mechanical movement with manual winding.
Design for a "Pasha" wristwatch in platinum and diamonds, presented at SIHH, Geneva, in 1995. This model is set with 116 pavé diamonds with a total of about 2 carats. It has an automatic movement.

Opposite left: *"Santos galbée" bracelet watch, with automatic mechanical movement, Cartier, 1987*

Opposite right: *Two models of "Mini-Panthère" bracelet watch in platinum and diamonds. Cartier, 1993 The model at the left has 268 baguette-cut diamonds and 52 brilliant-cut diamonds.*

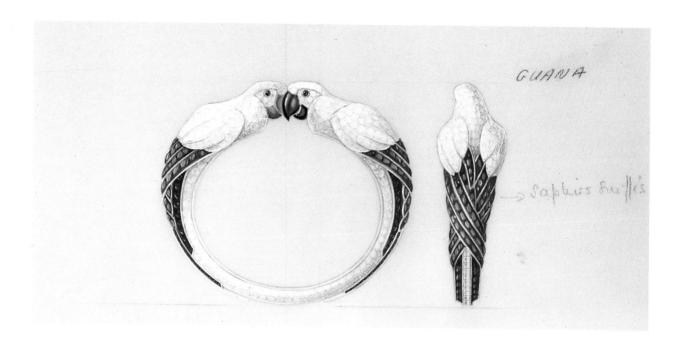

These two pages illustrate several designs by Cartier Joailliers (1995), now under study, which eventually will be realized as a new collection. The great tradition of Charles Jacqueau, Jeanne Toussaint, and the other legendary creators and designers of Cartier Paris is carried on today under the direction of Micheline Kanouï, in the painstaking work and elegant solutions of the team responsible for the creation of fine jewelry.

Above: *"Guana" bracelet*
Two macaws join head to head. Their bodies are covered with diamonds on platinum, their tail feathers are lines of tallow-cut sapphires, their beaks in gray mother-of-pearl and eyes in round emeralds.

Below: *Table clock with stand*
The cut-cornered square clock is rimmed with diamonds on platinum on the face and sides. The front is covered with calibré-cut sapphires and shaped sapphires. The off-center mystery dial echoes the shape of the case.

Opposite: *"Topkapi II" necklace*
The Indian-inspired necklace is made of diamonds with ribbed emerald beads studded with a diamond and briolette sapphires. Six suspended briolette sapphires flank the central pear-shaped emerald engraved in squares.

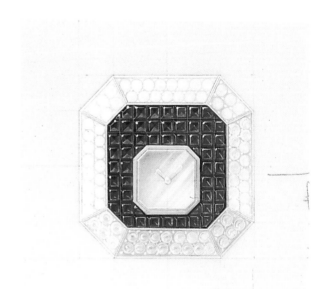

peut être à
supprimer.

HP700094

26646

CV.4

resserrer les parties Bts
pour libérer les Briolettes
à mettre au devis.

— caché avec la toilette
Travail style Coopérative

*T*he Cartier Foundation of Contemporary Art,
Paris (architect: Jean Nouvel, 1992)
For many years, French architect Jean Nouvel has
conceived his projects as true works of contemporary art, in which the stripped-down look of
metal, revealing its essential nature, is wedded to
glass.

CHRONOLOGY
PLATINUM

1551–1306 B.C.	Platinum, mixed with gold, appears for the first time in several jewels belonging to the Pharaohs of the Eighteenth Dynasty.
720 B.C.	One of the hieroglyphics that decorate a casket that belonged to the great priestess Shepenupet I, daughter of the king of Thebes, is in platinum.
100 B.C.	In an area corresponding to present-day Esmeraldas, Ecuador, the Pre-Columbian Indians made platinum jewelry with a technical prowess that remains astonishing today.
70 B.C.	In his *Historia Naturalis*, Pliny mentions a metal he calls *plumbum album*, or white lead, which some specialists believe could refer to platinum.
1520	Along the banks of the Pinto River in the Spanish colony of New Granada (present-day Colombia), the Conquistadors discover an unknown metal that they call *platina*, or little silver.
1557	The Italian scholar Giulio Cesare Scaligero publishes his *Exotericarum Exercitationum Liber*, the first scientific text that mentions platinum.
1679	A Czech Jesuit, B. Balbinus, simplistically describes platinum as white gold.
1707	A Spanish edict explicitly outlaws the use of platinum to adulterate gold.
1741	The first sample of platinum arrives in Europe, reaching England by way of Jamaica.
1748	The Spanish explorers Antonio de Ulloa and Jorge Juan publish an account of their expedition, which contains the first precise description of platinum and the places where it is found.
1751	English scientists Charles Wood and William Brownrigg present a report to the Royal Society on the properties of the new metal. Another scientist, the Swede Teophil Scheffer, is the first to dissolve platinum, using small quantities of arsenic.
1753	The Spanish government requests Irish naturalist William Bowles to study the new metal thoroughly and report on its possible uses.
1755	An anonymous publication entitled *Le Platine, l'or blanc et le huitième métal* (Platinum, White Gold and the Eighth Metal) appears in France.
1758	Giovanni Giacomo Casanova and the marquise d'Urfé carry out mysterious experiments, attempting to turn platinum into gold.
1759	Charles III, king of Spain, directs the Real Sociedad to study platinum.
1779	The Swede Torbern Bergman gives the new metal its scientific Latin name, *platinum*.
1780	French goldsmith Marc Étienne Janety creates the first precious objects in platinum, for King Louis XVI.
1782	French scientist Antoine Laurent de Lavoisier achieves the complete melting of platinum by enriching the flame with oxygen.

1784	German chemist F. K. Achard makes the first platinum crucible.
1789	Charles III, king of Spain, offers Pope Pius VI a platinum chalice that weighs four and a half pounds.
1790	The Barnabite Father Angelo Maria Cortenovis publishes the treatise *Du Platine connu des Anciens* (On Platinum Known by the Ancients).
1795	The metric system is adopted in France, and the official weights and measures of the meter and the kilogram are created in platinum. Josiah Wedgwood uses platinum to decorate his porcelains.
1802	Charles IV of Spain has a Platinum Room made in the royal palace at Aranjuez.
1802–4	Two English scientists, William Wollaston and Smithson Tennant, discover four other metals of the platinum group: palladium, rhodium, iridium, and osmium.
1813	Johann Wolfgang Döbereiner exploits platinum's properties to invent the lighter.
1819	Significant deposits of platinum are discovered in the Ural Mountains.
1820	Marie-Louise of Austria, duchess of Parma, wears ceremonial gowns embroidered with platinum threads.
1828	Czar Nicholas I of Russia orders that rubles in platinum be made a part of the legal currency.
1837	Samuel Morse invents the telegraph, using platinum for the electrical contacts.
1844	Russian scientist Carl Claus discovers the remaining metal of the platinum group, ruthenium.
1853	Cartier makes its first jewel (shirt buttons) in platinum.
1862	At London's International Exposition, a platinum ingot weighing over two hundred pounds, obtained from a single melt, is shown.
1866	Significant deposits of platinum are discovered at Kimberley, South Africa.
1873	Platinum is used in photography to make particularly sensitive prints rich in halftones.
1878	Joseph Swan makes the first incandescent light bulb using platinum for the filaments.
1888	Platinum is discovered in Canada.
1901	Nobel Prize–winning German chemist Friedrich Wilhelm Ostwald invents a process for the preparation of nitric acid using platinum as a catalyst.
1906	Cartier foreshadows the Art Deco style in jewelry, whose abstract lines and geometric shapes are admirably expressed in platinum.
1908	The Great Star of Africa, a diamond weighing 530.2 carats, is set in platinum for the British royal scepter.
1912	A German goldsmith from Pforzheim invents white gold as a substitute for platinum, which has become increasingly sought after for use in jewelry.

1924	Geologist Hans Merensky locates the largest deposit of platinum ever found, near Johannesburg, and South Africa breaks out in "platinum fever."
1927	American actress Jean Harlow becomes famous as the "platinum blonde," the inspired sobriquet originating with her producer, Howard Hughes.
1931	The platinum market collapses due to the Depression and overproduction.
1933	The platinum market stabilizes and prices climb up again.
1934	Platinum is used in the production of telescopes.
1937	The Duke of Windsor, ex-king of England Edward VIII, marries Wallis Simpson, and Cartier makes their wedding rings and bracelet celebrating the stages of their love affair in platinum.
1940	The United States government declares platinum a strategic material and forbids its use for civilian purposes.
1946	Almost all the platinum used for jewelry is bought by the American market.
1950	Platinum is used to refine gasoline.
1954	South Africa becomes the largest world producer of platinum.
1960	Among the growing industrial uses of platinum is a powerful anticancer medicine based on platinum.
1967	Elvis Presley gets married and gives his new wife a platinum wedding ring set with diamonds.
1970	International monetary crises lead to a notable rise in platinum production, which reaches three million ounces a year. The Clean Air Act in the United States gives birth to an industry producing catalytic converters, of which platinum is a crucial element. Japan experiences a platinum boom in the realm of jewelry.
1980	The price of platinum attains a record of $870 per ounce.
1982	Platinum jewelry becomes popular in Europe once again, especially in Germany and Italy.
1983	The Isle of Man strikes the first modern platinum coin of legal tender.
1986	Economic sanctions against South Africa create a certain instability in the platinum market.
1987	A new platinum mine is opened in Stillwater, Montana.
1990	World production of platinum reaches a record level of 4.3 million ounces, of which 77 percent comes from South Africa.
	American jewelers begin to offer more jewelry in platinum.
1995	The price of platinum averages between $410 and $420 per ounce, while gold sells for about $380. The market fluctuations of the two precious metals are generally linked together. The Japanese market remains the most important for platinum jewelry, which uses 46 tons a year. However, platinum makes a strong comeback in Europe and America.

CHRONOLOGY
CARTIER

1847	Louis-François Cartier buys the jewelry workshop at 31, rue Montorgueil, from his mentor Adolphe Picard.
1853	Cartier cultivates a private clientele, moving to 5, rue Neuve-des-Petits-Champs. At this time, Cartier's first object in platinum is made.
1856	Princesse Mathilde, cousin of Emperor Napoleon III, makes her first purchase.
1859	Cartier moves to 9, boulevard des Italiens. Empress Eugénie becomes one of its clients.
1860	Cartier has its first Russian client.
1872	Louis-François Cartier makes his son Alfred a partner.
1874	Alfred Cartier takes over the management of the store.
1888	The first bracelet watches are offered.
1898	Alfred makes his oldest son, Louis, his partner, renaming the business Alfred Cartier & Fils.
1899	Cartier moves to 13, rue de la Paix.
1902	Cartier opens a London branch at 4 New Burlington Street.
1904	The first appointment to the Court of England comes from Edward VII. Louis-François Cartier dies. Pierre Cartier travels to Russia. Cartier is appointed purveyor to the king of Spain, Alphonse XIII. Louis Cartier makes the company's first wristwatch with leather strap for his Brazilian friend and aviation pioneer, Alberto Santos-Dumont. Queen Alexandra of England buys a *résille* necklace in platinum and diamonds.
1905	Appointed purveyor to King Charles I of Portugal.
1906	Jacques Cartier takes over the London store. Louis and Pierre Cartier become partners in Cartier Frères. The first Art Deco abstract and geometric jewels are made. "Tonneau" watch with leather strap is created.
1907	Pierre Cartier goes to New York. Cartier presents its first exhibition in Saint Petersburg, at the Hotel Europe. Cartier appointed purveyor to Czar Nicholas II of Russia. Edmond Jaeger signs a contract with Cartier to supply it with fine watches.
1908	Appointed purveyor to king of Siam, Rama V (Paraminda Maha Chulalongkorn).
1909	Cartier London moves to 175–76 New Bond Street. Pierre Cartier opens a branch in New York at 712 Fifth Avenue. Patent taken out for the deployant buckle. Charles Jacqueau joins Cartier.
1910	Louis Cartier visits Saint Petersburg. Pierre Cartier sells the blue Hope diamond, bought the previous year from his brother Louis at a Parisian auction. The logo of two linked Cs makes its first appearance.
1911	The "Santos-Dumont" wristwatch is produced on a commercial scale. Jacques Cartier travels to the Persian Gulf. Louis Cartier and Charles Jacqueau travel to Moscow and Kiev.

1912	The Paris City Council offers Cartier's imperial egg to Czar Nicholas II (today owned by The Metropolitan Museum of Art, New York). The first baguette-cut diamonds appear. "Comet" table clocks are made.
1913	The first mystery clock, "Model A," appears. Cartier is appointed purveyor to King Peter of Serbia.
1914	The panther decorative motif makes its first appearance, in diamonds and onyx, on a watch. Cartier is appointed purveyor to Philippe, duc d'Orléans.
1917	Cartier New York moves to 653 Fifth Avenue, the Morton F. Plant mansion, obtained in exchange for a necklace with two strands of 55 and 73 pearls.
1918	Cartier makes field marshal's batons for marshals Foch and Pétain.
1919	The "Tank" watch is created. Founding of the European Watch & Clock Co., Inc. Cartier is appointed purveyor to King Albert I of Belgium.
1921	Appointed purveyor to the Prince of Wales. Cartier Frères becomes Cartier SA.
1923	The first portico mystery clock, "Billiken," appears.
1924	Creation of the Rolling Ring three-gold motif in a ring and a bracelet.
1925	Alfred Cartier dies. Cartier participates in the Exposition Internationale des Arts Décoratifs et Industriels Modernes in Paris, exhibiting not with the jewelers but with the great couturiers in the Pavillon de l'Élégance.
1926	Cartier opens a branch in Saint-Moritz (it closes in 1945). The jeweler is appointed purveyor to King Fuad I of Egypt.
1931	Cartier makes its first sword for an academician of the Académie Française, the duc de Gramont. A mystery pocket watch is made.
1933	Jeanne Toussaint is named head of the fine jewelry department. Cartier takes out a patent on its invisible setting (*monture invisible* or *serti mystérieux*).
1935	Cartier opens a shop in Monte Carlo.
1938	Cartier opens a shop in Cannes. The smallest bracelet watch in the world, signed Cartier, is offered to Princess Elizabeth of England.
1939	Cartier is appointed purveyor to King Zog I of Albania.
1940	Paris is occupied by German troops. Charles de Gaulle establishes the Free French Forces in London. His BBC broadcasts are made at the London offices of Jacques Cartier.
1941	Cartier creates the jewel known as the caged bird, symbolizing the German occupation. (After France's liberation, it was followed by the "freed bird.")
1942	Louis Cartier dies, followed by Jacques Cartier.
1945	Pierre Cartier becomes head of Cartier Paris and Cartier New York. Jean-Jacques Cartier, son of Jacques, directs Cartier London.

1948	Claude Cartier, son of Louis, takes over Cartier New York. The Duchess of Windsor orders a panther brooch from Cartier (bought back for the Cartier Collection at an auction held by Sotheby's in Geneva in 1987).
1955	Cartier makes a sword for academician Jean Cocteau from a sketch by the writer himself.
1965	Pierre Cartier dies.
1968	A luxury cigarette lighter carries the name Cartier.
1969	Cartier buys at auction a pear-shaped diamond of 69.42 carats and sells it to Richard Burton for Elizabeth Taylor. Alain-Dominique Perrin joins the company Briquet Cartier.
1970	Cartier Hong Kong opens.
1971	Cartier Munich opens.
1972	A group of investors headed by Joseph Kanouï buys Cartier Paris. Robert Hocq, who created the cigarette lighter of 1968, becomes president. A new collection of watches comes out, to be sold in all the Cartier stores as well as through a network of selected dealers.
1973	Robert Hocq, with the assistance of Alain-Dominique Perrin, creates the concept of Les Must de Cartier. Perrin is named its director. The first Les Must de Cartier boutiques open in Biarritz and Singapore.
1974	The same group of investors buys Cartier London. The first collection of fine leather goods is launched. A Les Must de Cartier boutique opens in Tokyo.
1976	The first collection of Les Must de Cartier watches in vermeil. The first oval pen comes out.
1978	The "Santos" watch in gold and steel is introduced. The first Cartier tie collection is offered.
1979	Cartier's international interests are brought together in the newly created Cartier Monde, which reunites Cartier Paris, Cartier London, and Cartier New York. Robert Hocq dies as the result of an accident. Joseph Kanouï is named president of Cartier Monde.
1981	Cartier SA and Les Must de Cartier SA merge. Alain-Dominique Perrin becomes president of Cartier SA and Cartier International. The "Must" and "Santos" perfumes are introduced.
1982	Micheline Kanouï becomes the creative director of fine jewelry and launches her first collection of "Nouvelle Joaillerie," or New Jewelry, on the theme of cabochons in precious stones.
1983	The concept of the Cartier Collection, a showcase of Cartier's heritage, is born. The "Panther" watch is introduced, as is the "Must" and "Vendôme" sunglasses.
1984	The Cartier Foundation of Contemporary Art is inaugurated. The second collection of New Jewelry, inspired by pastel-colored sapphires, comes out. Hans Nadelhoffer's book *Cartier: Jewelers Extraordinary* is published.
1985	The "Pasha" watch is introduced.
1986	The third collection of New Jewelry comes out, based on animal themes and pearls. The "21" watch in steel is created, as is the "Pasha" pen.

1987	The "Panther" perfume is introduced, as is the "Arts de la table" collection, "Les Maisons de Cartier" (porcelain, crystal, silverware).
1988	The fourth New Jewelry collection, on the theme of Egypt, appears. "Personal Line," accessories for men, is introduced. Cartier inaugurates its new international headquarters on rue François Ier in Paris.
1989	The book *Le Temps de Cartier* is published.
1989–90	The exhibition "L'Art de Cartier" is held at the Musée du Petit Palais in Paris.
1990	The "Panther" line of handbags is introduced.
1991	The fifth New Jewelry collection, on the theme "On the Road to the East Indies," appears. The "Panther" pen and the "Diabolo" watch are introduced. The first Salon International de la Haute Horlogerie (SIHH) is held at Palexpo in Geneva, where the six Art Deco portico table clocks, reunited for the first time, are exhibited.
1992	At the second Salon International de la Haute Horlogerie, the new collections of "Baignoire," "Casque d'Or," and "Belle Époque" watches are presented. The "Mini-Panthère" watch is introduced. The exhibition "L'Art de Cartier" is shown at the Hermitage in Saint Petersburg. The "Cougar" pen is introduced. The book *Made by Cartier* is published. A sword for the academician Jean-François Deniau is made.
1993	The third Salon International de la Haute Horlogerie is held in Geneva. The Chronoreflex watches ("Pasha," "Cougar," and "Diabolo") are introduced. The perfume "Must II," the "Panther" minihandbags in crocodile and pastel colors are introduced. A new watch, "Must II," and the ring "Ellipse" are introduced. The new Vendôme Luxury Group is created on October 23, which embraces Cartier, Piaget, Baume & Mercier, Dunhill, Montblanc, Chloé, Lagerfeld, Sulka, Hackett, and Seeger.
1994	SA and the Cartier Foundation inaugurate their new headquarters, the work of architect Jean Nouvel. The black and gold leather collection is launched. The fourth Salon International de la Haute Horlogerie is held in Geneva, and three new collections of watch-jewels are presented, on the themes "Art Deco," "Saint Petersburg," and "Pearls." A limited edition of the "Santos-Dumont" watch in platinum and pink gold is offered on its ninetieth anniversary.
1995	The fifth Salon International de la Haute Horlogerie is held in Geneva. On the tenth anniversary of the "Pasha" watch, a new "Pasha" is presented. The new Louis Cartier pen is introduced. The exhibition "L'Art de Cartier" is mounted at the Teien Museum in Tokyo. Cartier presents the "MK" engagement ring in platinum and diamonds to Japan. Platinum has definitely entered everyday life.

BIBLIOGRAPHY

The basic bibiography given here is limited to the two major themes of this book: Cartier (its history, stylistic and technical achievements, production) and platinum (its discovery, extraction, physical and chemical properties, and uses). The enlargement of the subject to include the important movements of the twentieth century in the fields of culture, art, and ideas was the fruit of the authors' independent thinking and is not based on any original research carried out by other people, relying only on the simple consultation of manuals and catalogues.

Barracca, Jader; Negretti, Giampiero; and Nencini, Franco. *Le Temps de Cartier*. Paris: Michel de Maule, 1990; English-language edition: Milan, Wrist International, 1990.

Ceschina, D. "Il Platino et il suo impiego tra età moderna e contemporanea." Thesis, Università degli Studi di Milano, Facoltà di Scienze Politiche, 1989–90.

Cologni, Franco, and Mocchetti, Ettore. *L'Oggetto Cartier: 150 anni di tradizione e innovazione*. Milan: Giorgio Mondadori, 1992; Paris-Lausanne: Bibliothèque des Arts; English-language edition: *Made by Cartier*. Milan: Giorgio Mondadori, 1992.

Gautier, Gilberte. *La Saga dei Cartier*. Milan: Sperling & Kupfer, 1987; English-language edition: *Cartier the Legend*. London: Arlington Books, 1987.

McDonald, D., and Hunt, L. B. *A History of Platinum and Its Allied Metals*. London: Johnson Matthey, 1982.

Nadelhoffer, Hans. *Cartier*. Milan: Longanesi, 1984; English-language edition: *Cartier: Jewelers Extraordinary*. New York: Harry N. Abrams, Inc., 1984.

PHOTOGRAPH CREDITS

Archipress: pp. 182–83.

Archives Cartier Paris: pp. 11, 13, 15, 17, 18, 23, 25–27, 29, 30, 34, 35, 37–39, 41, 46–49, 51, 53, 54, 58, 60, 61, 63, 64, 67–71, 74, 75, 76a, 78, 79, 83, 84, 85, 87, 88, 90–97, 99, 100, 102b, 103, 116, 124, 138, 148a, 148b, 153, 204, 205a, 210, 211b, 214a, 214b, 216, 224a, 224c, 226, 228c, 228d, 230–45, 246a, 247–60, 262, 263, 264, 266, 267.

Archives Cartier London: pp. 36, 42, 146, 147, 149.

Archives Cartier New York: pp. 133b, 134, 135, 145b, 169b, 171, 172, 181b, 186b, 190–93, 198–200, 202, 203, 205b, 208, 209, 211a, 215, 246b.

Archives Platinum Guild International: pp. 7, 8, 19, 20, 31, 32, 43, 44, 55, 56.

Cartier of H. Nadelhoffer: pp. 139a, 151, 179.

Christie's: pp. 101, 120, 121, 129a, 175, 184a, 184c, 186a, 188c, 189, 194a, 219a, 219b, 219c, 220b, 224b.

Collection Cartier: pp. 76b, 77b, 81, 82, 86, 89, 102a, 104, 105a, 105b, 105d, 106, 107, 110–15, 117–19, 122, 123, 125, 126, 128, 129b, 129c, 130–32, 133a, 136, 137, 139b, 140, 141, 144, 145a, 148b[credited to Cartier Paris], 150, 152, 154–57, 159–63, 166–68, 169a, 170, 173, 178, 181a, 188a, 194b, 194c, 195, 196b, 197, 201a, 206a, 207, 211c, 213, 214c, 217, 218.

Electa: p. 228a, 228b.

Fondation Cartier: pp. 268–69.

Phillips: p. 196a.

Private collection: pp. 164, 165.

Sandro Sciacca: pp. 227, 229, 261, 265.

Sotheby's: pp. 174, 176, 177, 185, 187, 188b, 188d, 201b, 206b, 206d, 219d, 219e, 220a, 220c, 220d, 221, 225.

Louis Tirilly: pp. 76c, 80, 98, 105c, 127, 180, 206c, 212.

Roger Viollet: pp. 10, 22, 72–73, 108–9, 142–43, 222–23.